HOW LIGHT
MAKES LIFE

How Light Makes Life: *The Hidden Wonders and World-Saving Powers of Photosynthesis*
Copyright © 2021 by Raffael Jovine

Originally published in the UK as *Light to Life: The Miracle of Photosynthesis and How It Can Save the Planet* by Short Books in 2021. First published in North America in revised form by The Experiment, LLC, in 2022.

The Experiment, LLC
220 East 23rd Street, Suite 600
New York, NY 10010-4658
theexperimentpublishing.com

THE EXPERIMENT and its colophon are registered trademarks of The Experiment, LLC. Many of the designations used by manufacturers and sellers to distinguish their products are claimed as trademarks. Where those designations appear in this book and The Experiment was aware of a trademark claim, the designations have been capitalized.

The Experiment's books are available at special discounts when purchased in bulk for premiums and sales promotions as well as for fundraising or educational use. For details, contact us at info@theexperimentpublishing.com.

Library of Congress Cataloging-in-Publication Data available upon request

ISBN 978-1-61519-863-4
Ebook ISBN 978-1-61519-841-1

Cover and text design by Jack Dunnington
Cover photograph by Cristian M. Vela/Alamy Stock Photo
Illustrations by Nora Jovine

Manufactured in the United States of America

First printing June 2022
10 9 8 7 6 5 4 3 2 1

HOW LIGHT MAKES LIFE

THE HIDDEN WONDERS AND WORLD-SAVING POWERS OF PHOTOSYNTHESIS

RAFFAEL JOVINE

THE EXPERIMENT

NEW YORK

CONTENTS

PROLOGUE

"I remember in college another guy and I had an idea that—mind if I talk about myself?"

"If you don't, I will."

"Well, this guy and I had this idea. We wanted to find out what made the grass grow green. That sounds silly and everything but the biggest research problem in the world is there. I tell you why. Because there is a tiny little green engine in the green of this grass and in the green of the trees that has a mysterious gift of being able to take energy from the rays of the sun and store it up. You see that's how the heat and power in coal and oil and wood is stored up. We thought that if we could find the secret of all those millions of little engines in this green stuff, we can make big ones. And then, we can take all the power we can ever need right from the sunrays, you see . . ."

"That's wonderful, I never knew that."

"We worked on that, worked on, day and night, we got so excited we forgot to sleep. If we

made just one little discovery, we would walk on air for days . . ."

"Yes? And then what?"

"Well, then we left school. Now he is selling automobiles and I'm in some strange thing called banking."

—Jean Arthur and James Stewart, *You Can't Take It with You* (1938), Columbia Pictures

My long relationship with photosynthesis began with this 1938 black-and-white romantic comedy, in which a love-struck banker's son convinces a brilliantly eccentric academic's daughter to marry him. The pivotal moment is when the banker's son finally reveals his abandoned passion for research into the hidden power of plants.

Until I saw this movie, I had been a sickly kid who worked as a child actor instead of attending school. I was mesmerized. "That green engine thing makes so much sense!" I shouted in my head. I concluded if that is what it takes to get the girl, I needed to know everything about this enigmatic thing called photosynthesis.

I stopped skipping school and, with the help of an inspired teacher, I learned how to build a greenhouse. My acting career was over, and I became the first conventional high-school graduate in my bohemian family. Studying photosynthesis took me from being an actor to completing post-doctorate studies in molecular biophysics and biochemistry.

You Can't Take It with You is not really a film about plants and science. Most people watching it would say it explores whether material wealth provides emotional satisfaction. Fortunately, in the eighty-two years since it was made, much more progress has been made in photosynthesis research than in answering questions about what makes us happy.

I still want to store the power of sunrays. Unfortunately, studying photosynthesis does not always get you the girl. Fortunately, I did not know that at the time.

Why write about photosynthesis?

Photosynthesis is the conversion of sunlight into chemical energy, which involves an indescribably intricate collection of biological processes. This book does not explore this beautiful and complex machinery. Instead, it looks at what the value and the impact of this "green stuff" is, for people and for the planet. This book is not intended for teaching purposes. It definitely will not suffice as reference material and it will skip over vast bodies of brilliant research. But it can, I hope, communicate at least some of that tingle of glee and wonderment scientists feel when we get the chance to reveal a tiny facet of the magic of nature.

It is a process that, as so often with biology, is described as if it was merely reacting to survive chemical, geological, or cosmic pressures. Such a view obscures how turning light into life has shaped our planet and continues to do so. Photosynthesis does not happen in spite of the foundational and powerful forces of nature; photosynthesis itself is one of these forces.

Almost from the beginning of life on Earth, photosynthesis has generally increased living spaces and vital resources. It has transformed our oceans, created our atmosphere, sculpted our mountains and continents. It has turned our once hyper-hostile young planet into a vibrant, elegant, and abundant world into which it is still pumping more life energy and usable energy than anything else does. At the same time, it has managed to be gorgeous, colorful, and surprising.

Our own species, one that shares about 40 percent of our genes with apples, is very much a direct product of this process. The deliberate manipulation of photosynthesis has largely been responsible for the mind-boggling success of *Homo sapiens*. Today, amid all the challenges we face, photosynthesis remains a dominant force large enough, fast enough, and powerful enough to rebalance our complicated world, even after hundreds of years of human mismanagement. It is the most affordable and attainable way to repair our exhausted ecosystems before we run out of time. And it will do all this while feeding us, providing us with new jobs, and improving habitats for all creatures great and small.

But only if we let it.

The exploitation of our natural environment in ugly and counterproductive ways is continuing to accelerate. The developed world wants to protect the enormous gains it has created in health and wealth by gobbling up the last remaining resources. Developing nations want to use more energy and consume even more of the same resources to catch up.

Across the world, policy makers and business, community and spiritual leaders are beginning to experiment with carbon taxes or similar disincentives to reduce carbon emissions. We are constantly reminded that too much carbon dioxide is destabilizing our weather and responsible for everything from floods to famine.

Such efforts to reduce carbon use are clearly necessary but they are also insufficient. Without the natural removal of emissions from the environment, emissions will continue to accumulate. We need to not only reduce emissions but also increase the planetary capacity to regenerate and refresh the atmosphere. In practice, there is only one way to do this.

Amid all the noisy and bitter arguments over weather, photosynthesis has spent four billion years quietly broadcasting a constructive, consistent, and positive message every day on every continent and across every ocean. You can taste, smell, touch, and see the benefits of photosynthesis every time you eat and with every breath you take.

Photosynthesis is the original, largest, and most powerful "Net Negative" technology ever, and it is at our fingertips right now. Increasing photosynthesis is something that every one of us can do, no matter where we live or what our income bracket. It works everywhere in the world and, compared to any other carbon reduction measures, it is extremely affordable.

In this book, I hope to show that we have everything we need to stop global degradation while continuing to succeed as a species and develop our economies. Rebuilding our world is surprisingly easy to do and good for our health as well. Every seed carries the instructions to save our world.

The book is structured in seven chapters.

The first tells the story of how photosynthesis was discovered. It is stranger than you might expect because, for a long time, we did not understand what was staring us in the face. Gradually, over the course of four hundred years, human ingenuity deciphered the riddle of how sunlight turns inorganic matter into food. In doing so, we transformed our limited and pessimistic relationship with nature to one where we can feed the world, even with an enormous population.

Chapter 2 outlines our planet's relationship with photosynthesis. It shows how sun-fueled biological processes have made the world we know today and why biology is more than powerful enough to transform it again, several times over.

Chapter 3 asks who—or what—is actually responsible for this magic. The answer to this is full of surprises, with new organisms that have global impact still being discovered today. The richest resources and environments on our planet are created by a collaboration of photosynthesizing plants and animals, including ourselves.

Chapter 4 looks at different forms of photosynthesis across oceans and on land. It challenges preconceptions about how best we can absorb carbon and rebalance our planet's fragile systems. There are enormous opportunities, on every continent, in every environment.

In chapter 5, I posit reasons to be both alarmed and hopeful. The "good news" emphasizes the immense hurdles our species has overcome before and shows why we can yet find sustainable solutions to the huge challenges ahead. The "bad news" sets out just how far and how fast our excesses are unraveling hundreds of millions of years of biological progress.

Chapter 6 seeks to put a value on photosynthesis. Almost everything worth anything in our economies is either directly or indirectly made possible because of the process of harnessing sunlight energy, which is provided to us absolutely free. And, if we can show how valuable it is, perhaps people might appreciate it more.

And finally, in chapter 7, I ask where we can go from here. We have a choice. We can actively grow our planet again. We can produce more food, sustain more life, and create a fairer world. And, yes, we can do this while making more money, too. Best of all, each one of us, no matter our circumstances, can make a constructive, life-affirming contribution.

Scientists are humans and, as such, are sometimes driven by the desire to get rich or squeeze more profit from nature. More often we are simply inspired by our love for the weird, wonderful, dazzling organisms that we study. And all of us hope that we may have the chance to make the world a bit better.

I know this to be true because those same motivations have driven me. I started out by building and planting a greenhouse in secondary school because I loved watching things grow. Even then I wanted to be a research biologist mostly because it was the most exotic work I could imagine, so much more interesting than the boring films my family made. Those were the days when the new discipline of genetic engineering was all over the news and I wanted to tailor organisms like Eldon Tyrell did in

Blade Runner; I wanted to grow new eyes for my blind brother.

In college, I learned the first DNA sequencing methods and how to manipulate genes in yeast, and how to crystallize proteins. After graduating in molecular biochemistry and biophysics, I took a job in a research lab at a time when chemical photosynthesis was all the rage, and I tried to help decode how light energy is converted into chemical energy so that we could make biologically inspired solar-electricity-generating cells.

But my first piece of fieldwork, looking at marine photosynthesis, made me realize that months in laser laboratories had not prepared me sufficiently to learn about its true potential. Organisms in the real world are more sophisticated and resilient than I had ever imagined. Watching the response of the algae that lived under the ice in Antarctica to the Ozone Hole demonstrated that natural communities were flexible, responsive, and tougher than I would have believed as a molecular biologist. When the conditions are just right for them, many algae can quickly acclimatize to their environment and divide so fast that they outgrow everything around them. I began studying what triggered algal "blooms" so that we might be able to prevent and better manage so-called red tides, which can disrupt ecological systems.

Then life interfered again in the form of a growing

family and a recession. I became a management consultant and, just as interest in algal biofuels began to intensify, found myself working on balance sheets to help merge pharmaceutical companies.

Stuck in the office, I dreamed of field work. I missed algae. When my biologist friends complained it was too expensive to grow algae and told me that the US Navy was paying $700 per gallon of biofuel, I was puzzled. Algae had been growing on a truly enormous scale for free, like clockwork, every spring and autumn, for three and a half billion years. Puddles automatically fill with algae. Which made me ask, "Exactly what is so difficult about growing algae?"

Using my newfound consultant skills, I evaluated what made growing them so expensive. Brilliant engineers had attempted to grow known laboratory strains of algae to an industrial scale in artificial conditions. The results were clever but the systems were complex and brittle with enormous costs. At every stage, the objective was to control the organisms both inside and outside the reactors, whereas in nature, the organisms control their environment. In an effort to copy nature, I started looking at how to apply the "bloom" formation process found offshore to growing algae in manmade ponds. I designed and patented a process that replaced the expensive heavily engineered methods with low-cost natural ones.

Given population and environmental trends, I knew

we would need more food. I hoped I could address an imbalance and started a company. Soon we were testing the idea at an abalone farm in South Africa. Next was a research project in Oman that was 260 times larger, followed by a site 15 times larger than that in the Sahara Desert in Morocco. Unsurprisingly, working with nature is challenging, but we are now growing microalgae to produce animal feed, human food, high-value chemicals, pigments, and pharmaceuticals. We are creating a new very valuable food out of desert, seawater, and sunlight. We are increasing the size of our production system and, in our small way, doing a bit to help rebalance the planet—by using the oldest growth system on Earth.

Although we live in a troubled world, we can all explore new opportunities to make a better and richer one.

Time and time again our species has overcome seemingly impossible challenges. We can do so again. And, best of all, every one of us can contribute in a practical way—without having to go to the desert to grow algae—while having a lot of fun.

CHAPTER 1

WHAT IS PHOTOSYNTHESIS?

When my wife wants to get my attention with an urgent email, she writes "Algae, Algae, Algae" in the subject line. Embarrassingly, I fall for this trick every time. However, I will reluctantly concede that not everyone is as fascinated by single-celled organisms as I am.

And I will also recognize that many readers may be forgivably hazy as to what this weird fourteen-letter word "photosynthesis" means. Some may even have bad childhood memories of trying to learn the following formula:

$$6CO_2 + 6H_2O = C_6H_{12}O_6 + 6O_2$$

So, at the outset of this book, let me try to describe this miraculous energy machine.

1. Photosynthesis is the process by which biological organisms capture light energy to combine simple environmental chemicals like water and gases and convert them into useful and more complex compounds that fuel growth, reproduction, movement, and almost all life on this planet.

2. Organisms that use such external energy sources are called primary producers. These form the base for all food chains, including our own.

3. Photosynthesizing organisms are not just green land plants. There are many others that are very important for balancing global nutrient, energy, and chemical cycles.

Discovery

Until four hundred years ago, the conventional wisdom was that plants "ate" dirt. Yes, we believed those green rascals chomped on the soil. People held this view despite tens of thousands of years in which they had tended, observed, selected, named, classified, and processed plants with the greatest care and sensitivity.

We had discovered that fruit, grains, nuts, and berries make highly nutritious food, and storing it for lean times obviously made sense even to stone-age hunter-gatherers.

Planting crops enabled people in lush and meteorologically benign places to settle. Producing, trading, and preparing food became—as it still is today—a major source of wealth and employment across the world. Our clothing, housing, and very survival depended on growing plants.

Throughout all this time, humans—the smartest creature on the planet—had prospered by learning from their environment and noticing the smallest of details. And still, somehow, for several millennia, people did not spot that plants need light. Instead, the prevailing dogma was that plants consume the soil that they grow on. I suppose it is possible to see how such a misconception could have started. Fallow fields, which have had a season to recuperate after a harvest, do better than those that are continuously farmed. Spent soils are spent. That is still true today, so maybe the misconception that plants eat soil was related to our understanding of the need to recharge the land.

Even so, it is still pretty remarkable that we did not make a connection with light. The sprout on a potato that has been stored over winter in a light-tight basket is yellow. If you then take it out and expose it to light, the same sprout turns green within a day. Surely people observed such changes in coloration? You would have thought that farmers would have made the connection between light and the satisfying green look of plants; that Peruvian potato cultivators, who cherished their beautiful flowers, would have worked out how much light their favorite spuds needed. What about sunflowers that turn their faces to the sun? It seems improbable that nobody made the connection between growth and the sun, and yet it was not documented.

Mystics, priests, and physicians

It was not until the seventeenth century that a better understanding began to emerge but, frankly, even then it was faltering progress.

The first breakthrough came from a Flemish alchemist whose medical experiments, apparently inspired by dreams of the archangel Raphael, had to be hidden from the Spanish Inquisition. Jan Baptist van Helmont was a physician primarily focused on the study of human physiology. He introduced the mystical term "gas" into medical literature. He discovered that the "fermenting must"—carbon dioxide in today's language—made the air in wine cellars unbreathable and was carried in blood running through veins and expelled by the lungs.

In the 1620s, van Helmont turned his attention to testing the dogma that soil was consumed by plants. He cultivated a willow tree sapling in a covered pot, filled it with two hundred pounds of carefully weighed and dehydrated soil, then watched the tree grow. Five years later, he found the tree had gained 169 pounds in weight, while the soil weighed only two ounces less. He concluded, partially correctly, that plants consume water instead of soil. One can quibble with the exact methods, but this was the first modern reference to a structured and often repeated experiment that tested what plants need to grow.

Like so many discoveries in science, the idea for it had probably been around a lot longer. Leonardo da Vinci had conducted a similar experiment with pumpkin seeds that was later discovered in his unpublished notebooks.[1] Da Vinci, in turn, may have been inspired by Nicholas of Cusa, who suggested the experiment in his book *De staticis experimentalis* in 1450.[2] And Nicholas of Cusa himself may have been inspired by a Greek text dated between 200 and 400 CE called *Recognitions*. Ideas can hang around for thousands of years before someone actually tests them.

Van Helmont got us halfway there by proving that plants don't eat soil. Ironically for someone who had studied the biological function of carbon dioxide, he completely missed the fact that his tree was consuming gases and light. Perhaps, had the Spanish Inquisition not placed him under house arrest for blasphemy, his findings might have been debated and repeated sooner. Instead, they remained unnoticed until 1648, when his book *Ortus Medicinae*[3] was published posthumously by his son.

A further advance in human knowledge almost happened in 1679 when a French physicist and priest, Edme Mariotte, developed a theory that plants might acquire part of their nourishment from the air. He was a gifted man who was present at the foundation of the French L'Académie des Sciences, but unfortunately he failed to publish his findings. Indeed, *Discours de la nature de l'air, de la végétation des plantes. Nouvelle découverte touchant la vue* only made it into print 250 years later, in 1923.[4] Note to young scientists: Please publish, disseminate, test, repeat, and discuss your findings—do not expect your ideas to emerge by magic. And, even then, sometimes ideas have to circulate for a long time before they are accepted.

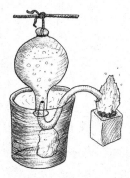

Figure 1: In 1727, Stephen Hales built his own equipment to trap gases, which could then be fed to plants. Image of combustion gas trap redrawn from Vegetable Staticks [*sic*].

Instead, we had to wait until the eighteenth century for the next landmark on this slow journey of scientific discovery. Stephen Hales, an English clergyman whose many achievements included the first measure of blood pressure, began studying a process called transpiration—or the loss of water from leaves. He surmised that "plants very probably draw through their leaves some part of their nourishment from the air."

Huffing and puffing through his lung-powered homemade bellows and strange "re-breathing" machines, he experimented with inverted bottles, and observed that the volume of air above the water surface decreased when a plant was grown in a closed atmosphere. He concluded that air was "being imbibed into the substance of the plant." In his 1727 book, *Vegetable Staticks,*[5] he even conjectured that light might be an energy source for plants.

These ideas began to percolate through Europe. In 1779, a Dutch physician called Jan Ingenhousz decided to test the idea that plants exchange air by submerging leaves under water in sealed bottles and then waiting for bubbles to form. Bottle after bottle of drowned leaves testified to the failure of these early experiments until a shaft of sunlight accidentally illuminated one of his bottles. Within minutes, he was watching the formation of the long-awaited bubbles.

Ingenhousz, anchored in the scientific tradition

of the day, was surprised that light had anything to do with it. He repeated the experiment with many different plants in dark and light bottles. He even tested whether it was thermal heating from his fireplace or the visible light of the sun that caused the bubbles to form. Eventually, he deduced that light was necessary for bubble formation and concluded the released gas was "fire air," soon to be called oxygen. He went on to test which gases were emitted in the dark and described these as "damaging the air" in contrast to those emitted during the day, which "purified" the air. Today we know that the plants removed carbon dioxide from spent air during the day.

By the end of the Enlightenment, we were really making progress, and scientific understanding of photosynthesis was about to take a giant step forward, thanks to Erasmus Darwin. He was a polymath physician, pathologist, and botanist, as well as the grandfather of the rather more famous Charles Darwin. He was also a slave-trade abolitionist, a supporter of women's education (especially for his illegitimate daughters), and the writer of bizarre poems such as "The Loves of the Plants."[6] It was in this poem in 1789 that he first touched on heavily concealed concepts of biological evolution.

This poem, found in Part II of Darwin's book *The Botanic Garden,* is prefaced with an "Apology" that outlines how it presents theories and scientific information through the deployment of mythic beings—gnomes, sylphs, nymphs, and salamanders, as well as deities of Egypt, Rome, and Greece—in the hope of making them more accessible. Without an index or any discernible structure, the poem merrily jumps from topic to topic. For example, the first canto contains the following order of subjects: "Hesperian Dragon, Electric Kiss. Halo around the heads of Saints, Electric Shock, Lightning from Clouds. Cupid snatches the Thunder-bolt from Jupiter, Phosphoric Acid and Vital Heat produced in the Blood. The Great Egg of Night. Western Wind unfettered." *The Botanic Garden* is made up of two books, each with four cantos, and the whole thing is a wild muddle, interspersed with additions, arguments, footnotes, and lists of scientific findings and data.

Yet, in "Note 5" on Sun Rays, Erasmus Darwin adds the following to Canto 1.I.136:

Some modern philosophers are of opinion that the sun is the great fountain from which the Earth and other planets derive all the phlogiston [burnable material] which they possess; and that this is formed by the combination of the solar rays with all opake [sic] bodies, but particularly with the leaves of vegetables, which they supposed to be organs adapted to absorb them. And that animals receive their nourishment from vegetables they also obtain in a secondary manner their phlogiston from the sun. And lastly as great masses of the mineral kingdom, which have been found in the thin crust of the earth which human labour has penetrated, have evidently been formed from the recrements of animal and vegetable bodies, these are also supposed thus to have derived their phlogiston from the sun.

The additional notes added to the poem continue with some wildly improbable theories about how the sun works. Yet, in his obscure way, Darwin had stumbled on exactly what happens with photosynthesis and plant growth.

Emerging from this melange of archangels, alchemy, sylphs, and experiments with homebuilt apparatuses was a new understanding that plants absorbed air and water to produce usable energy. Crucially, this early biology had also enabled people to see that sunlight was necessary for the process of turning water and gases into food and oxygen. This is the core tenet of photosynthesis. Although this new idea took hundreds of years to develop, it eventually removed the misapprehension under which human beings had labored for millennia: that plants ate soil.

There is another lesson in these stories. Even at a time when communications were desperately slow and books were being published posthumously by quirky investigators, ideas could still migrate across countries, grow, and trigger structured experimentation. Sure, the equipment was crude, messages were concealed in obscure analogies, and progress was painfully slow. Of course, many other scientists confounded the debate, with learned repetition of the status quo. Yet the lessons from these experiments were reproducible and demonstrable, and slowly they began to transform our understanding of nature.

Think for a moment about the danger of the old belief that land was being consumed by hungry green creatures called plants. This view made the world a finite, declining, and pessimistic place. In contrast, by learning that plants turn sunlight, water, and air into food, we discovered that the world was not finite—that there was "new" production. And that this plant-mediated production complemented our animal-based consumption: They inhale what we exhale. Instead of a world in which plants and animals were both eating up a limited amount of food and land on the planet, we were understanding that expansion and growth were possible. And, at the dawn of the nineteenth century, that was very necessary.

Feeding the masses, fueling the revolution

The sporadic scientific progress being made by alchemists, priests, poets, and physicians was about to accelerate. Revolutions—political, agricultural, and industrial—were creating modern nation-states with educated, organized, and potentially rebellious populations that demanded ever-greater quantities of food.

Three scientists, all of them very much products of their tumultuous century, and often a degree of personal torment, deserve credit for advancing our knowledge at this crucial time.

The first of them was Justus Liebig. Growing up in Darmstadt in the Rhineland, he experienced the famine of 1816—the "year without a summer"—when food crops were destroyed under the darkened skies caused by the explosion of the Tambora volcano in Indonesia. The experience is said to have shaped his later work as the founder of agrochemistry, a branch of science that would harness the power of plants to feed the world like never before.

His first experience of chemistry was helping his father mix paints and pigments for sale while experimenting with "fulminating mercury." Then bad things began happening to him. He was expelled from his pharmaceutical apprenticeship after burning down the attic of his mentor. Later he was forced to flee university because of his sympathies for nascent German nationalism. But Liebig got a second—and even a third—chance, possibly because he was a particularly beautiful young man. His sonnet-writing friend August Platen described him as a "lean figure, with an earnest friendliness in his symmetric face and large brown eyes with shadowy eyebrows, [which] arrest you in first seeing him."[7] He secured a grant to study at the Sorbonne in Paris with many of chemistry's greatest minds, including Alexander von Humboldt, who, impressed with Liebig's explosive mercury experiments, recommended him for a professorship at the University of Giessen. Some of the faculty there did not appreciate having a young professor with dodgy revolutionary ideas foisted on them and, initially, to make things worse, his fertilizer experiments failed. Ridiculed by his colleagues, struggling with funding, and sometimes paying for his staff and laboratory out of his own salary, he updated his book *Agrikulturchemie* seven times before it was published in its final form in 1856.

Finally, while struggling through multiple episodes of exhaustion, he succeeded in creating new nitrogen-based chemical fertilizers that would increase yields at reduced cost. He also demonstrated how depleted soils must be replenished with organic matter to support their health. He thus established for the first time a method and the tools by which farm yields could be improved on exhausted soils, without farmers having to leave the fields fallow while they regenerated naturally.

In the meantime, Liebig also co-discovered chloroform, found less toxic ways to make mirrors, revolutionized analytical chemistry, invented baking powder, created a meat extract to treat cholera patients that would later become Oxo cubes, and modernized baby food for orphaned babies. Of course, the university that once gave him such a hard time now carries his name.

The second key scientist in the nineteenth century was another German, Julius Sachs. He spent much of his childhood in Breslau going on expeditions with his father, an engraver, to collect samples of flowers and fungi that they would then paint. Sadly, while he was in his teens, his father, mother, and brother all died within a year of each other, and he was forced to earn a living by staining and preparing anatomical samples as teaching materials. However, he continued to pursue his botanic obsession, staying up all night—using vast amounts of cocaine—to meticulously categorize different plants.

Technological advances in microscopes, which enabled 300-to-600-fold magnification, combined with Sachs's sample preparation and tissue-staining skills, meant he could see live cellular structures in plants, including the units within cells called organelles and the "green granules," now known as chloroplasts, which are actually responsible for photosynthesis. By the time he was a famous university director, he had personalized high-powered microscopes and used them to paint beautiful large folios as visual aids for his lectures. These still rival modern photography in clarity and elegance.

In 1862, Sachs made a series of breakthroughs when he removed pigments from leaves and stained their chloroplasts with iodine. This is a substance that is selectively absorbed by starch, an energy-rich storage product made of glucose molecules. By comparing leaves that had been exposed to light with those that had been kept in the dark, Sachs discovered that the former produced starch, but the latter did not. Plausibly, he concluded that light was therefore necessary for starch production. Through his high-magnification observations of leaves, Sachs saw that their pores were actively opening and closing like our lungs do when we breathe. He understood not only that carbon dioxide was absorbed by the leaves when they were exposed to light and the pores were open, but also that this carbon dioxide was being converted into starch. And, because starch is made of the same glucose molecules as cellulose, Sachs then made the logical leap that starch was necessary to make the structural parts of plants such as stalks and stems.

At this point the baton was picked up by a third nineteenth-century pioneering scientist: Jean-Baptiste Boussingault. His career included two years working as a mining engineer and climbing volcanoes on the staff of the South American revolutionary General Simon Bolivar. On his return to his native France, he continued to dabble in politics. His views on decolonization and his role as a Republican member of the

French National Assembly caused him to be briefly stripped of his professorship in Paris. But his place in history is as a scientist. Boussingault was instrumental in quantifying the need for Liebig's nitrogen-based fertilizer for crop production and started the world's first agricultural research station. Most notably of all, in 1864, he made quantitative measurements of carbon dioxide uptake and oxygen production to balance the equation we are all taught when we first study photosynthesis:

$$6CO_2 + 6H_2O = C_6H_{12}O_6 + 6O_2$$

By 1893, all the components of the plant growth formula had been determined, whereby water and carbon dioxide in combination with sunlight are turned into sugar (glucose) and oxygen inside the green chloroplasts of plants. All the formula needed was a better name. Charles Barnes, the American botanist, declared that the terms "assimilation," "assimilation proper," and "assimilation of carbon" were inadequate or inappropriate.[8] He proposed the terms "photosyntax" and "photosynthesis" to describe the process of "synthesis of complex carbon compounds out of carbonic acid [carbon dioxide dissolved in water] in the presence of chlorophyll, under the action of light."

It had taken nearly three centuries for all these mystics, alchemists, priests, physicians, poets, cocaine-snorting scientists, and adventurers to describe the fundamental natural process of photosynthesis. With this new knowledge, people at last understood that plants inhale the carbon dioxide that we exhale, that they are a vital complement to animals, and that we are inextricably linked to plants through the shared medium of the air we breathe.

These insights bled across the borders of scientific disciplines. Toward the end of the nineteenth century, the idea that carbon dioxide and water vapor trap solar heat in the atmosphere was first conceived. Eunice Foote, an American women's rights campaigner, was the first to measure the heat absorption of different gases. She theorized that an atmosphere of carbon dioxide would heat the Earth. Three years later, John Tyndall worked out that the infrared portion of sunlight was absorbed by carbon dioxide, resulting in heat absorption.

Finally, this process was measured in our atmosphere and named by Svante August Arrhenius, Sweden's first Nobel Laureate. Using infrared observations of the light reflected from the moon, he determined: "If the quantity of carbonic acid increases in geometric progression, the augmentation of the temperature will increase nearly in arithmetic progression." He named the heat transfer from these atmospheric gases to the Earth's surface the "greenhouse effect."[9] Arrhenius used his calculations to gain a better understanding of how ice ages had been caused by photosynthesis absorbing too much carbon dioxide and thereby cooling the planet. He also reached the corollary conclusion that human carbon dioxide emissions would heat the planetary surface. Science has known about atmospheric heating by excess carbon dioxide since 1856, even if it has taken more than a century for people to begin considering whether we should do anything about it.

Let's pause for a moment and look at the scale of these achievements. The story of how we came to understand photosynthesis did not happen in isolation. It coincided with an enormous increase in the human population from 540 million to 1.65 billion.[10] There were three times as many mouths to feed and food production consequently needed to triple by creating better fertilizers and increasing agricultural yields. This could only have happened because we had discovered what made plants grow.

In the century or so since Charles Barnes gave the process a name, ever-better analytical techniques and modern chemistry have created a more nuanced understanding of photosynthesis. Scientists have found that green plants are not the only organisms that photosynthesize. They have discovered multiple different forms of photosynthesis, including weird microorganisms that do not actually make sugars.

These new discoveries have meant we have since adopted a broader definition of photosynthesis as "a process in which light is captured and stored by an organism, and the stored energy is used to drive energy-requiring cellular processes."[11]

Land plants are, of course, the photosynthesizers we know best because they are most easily reached and understood. But there are many more small organisms photosynthesizing, either living in extreme and unfamiliar places or hiding in plain view.

A great deal of the work needed to uncover these unfamiliar organisms has been done at the Hopkins Marine Station at Lover's Point in California. This research center was founded by the American railway tycoons Mary and Mark Hopkins, who picked a location where otters had been decimated for the value of their fur and there was an abundance of sea snails and other life in the ocean. Lover's Point got its name because Chinese fishermen, working at night with lights designed to lure squid, softly illuminated the beach for couples embarking on a nighttime stroll, romance, and more.

In the 1930s, a Dutch scientist, Cornelis Bernardus van Niel, whose previous work had been dedicated to isolating the bacteria that make the holes in Emmenthaler (Swiss) cheese,[12] joined the marine station. Traumatized by the First World War, he considered Lover's Point to be a relatively quiet retreat in which to cultivate "aesthetically pleasing" bacteria, and it was in doing this that he made the discovery that bacteria do not require carbon dioxide and water to photosynthesize like land plants do.

In a study of two particular types of bacteria, purple sulfur and green sulfur, van Niel carefully measured what they consumed and quantified their waste products, and discovered that both of these mud-dwelling bacteria use hydrogen sulfide—the source of the rotten-egg smell found in swamps—for photosynthesis. He found that other bacteria grew in the absence of oxygen but only if carbon dioxide and light were plentiful. He concluded that these forms of photosynthesis required something other than water as a reducing agent, and could grow on completely different nutrient sources such as hydrogen or organic compounds.

Thus van Niel set in motion the understanding that photosynthesis happens in completely different environments, where bacteria are supporting hugely diverse life-forms. In recent years, oceanographers have found entirely new photosynthesizing organisms that exist across the globe in great abundance. They have always been there, in every sea, but we simply did not see them.

During this time, another realization was dawning. Scientists worked out that the photosynthetic biomass—the organic material that comes from plants and animals—is the source of the coal, gas, oil, and dung that we use as fuel. That same biomass provides us with building materials such as wood or the mineral clinker for cement. And, of course, it feeds the fish, birds, and other animals we eat. Understanding that photosynthesis creates forests, food, and wealth from nothing but air, sun, water, and the Earth was—and is—a radical idea. It was as if the photosynthesizers were the real alchemists, turning base elements into life and wealth. It meant—and means—that the world is not finite.

In the four hundred years since van Helmont first weighed his tree, a deep transformation has occurred. Scientists are entirely human in their behavior and social constraints. They work in a political and emotional context just like everyone else. Sometimes their research is limited by equipment. Think of Hale huffing and puffing into his rebreathers. Others are simply lucky. Ingenhousz was not looking for light-generated gases when he accidentally let the sunshine on the leaves drowning in his bottles. Many of them were the kind of people who would never have been employed by research teams today.

So what? By collaborating, repeating, and discussing their findings, scientists test ideas. They document reproducible results and agree on the evidence, whether they are rivals or constrained by political or social norms. They do this through enormous upheavals, overcoming superstition and sometimes facing persecution. Some are superb theorists; others are excellent observers or brilliant experimentalists. Yet none of them work in isolation. By bickering and pointing out each other's mistakes, by arguing over the smallest details, they create something much larger than themselves, their academic disciplines, and the institutions they work in. They help us to understand the way the world works in spite of our beliefs. Without them, we would still treat the plague with bloodletting or mental health disorders with lobotomies. Without these compulsive specialists, we would still be using dog, donkey, and fly dung as therapeutics.

Every generation wants to create a better and healthier world for their children, and, in the face of huge challenges, curious minds help us to find new and better ways to do so. This is still true and still possible today.

CHAPTER 2

SUN-POWERED CHANGES TO EARTH

In the beginning

Earth formed 4.54 billion years ago from pieces of space debris at the swirling edge of a vortex that eventually became our sun. In this molten fireball, scraps of material—metallic rock, dust, and water—smashed into each other with such force that they all melted together to create our planet. Initially, the atmosphere was superheated steam and hot enough to render metals from rock. The heavier metals sank, forming Earth's magnetic core, while the minerals floated to the surface. Later, this witch's brew of chemicals cooled enough for the acidic atmosphere to condense into a corrosive rock-eating rain that continued for perhaps hundreds of millions of years until the first oceans formed.

The cooling was not all good news. The early sun was relatively small and less radiant, producing only about 70 percent of the light we see today.[13] Today, that would mean Earth would freeze. Fortunately, in addition to the carbon dioxide in the early atmosphere, there also was an estimated hundred parts per million of atmosphere-warming methane. That is more than fifty times the methane in our atmosphere today. Meanwhile, atmospheric sulfur and carbon chemically combined to form a new compound called carbonyl sulfide, which also trapped the heat of the faint sunlight so that the young Earth was kept warm and the oceans liquid under a blanket of carbon dioxide, methane, and carbonyl sulfide, which are all heat-absorbing gases. This steamy, brown, acidic world shrouded in noxious gases was barren and utterly hostile to life as we know it today. The oceans were filled with dissolved iron compounds that stained the water an opaque brown. Little light penetrated deep into the water.

And yet, in this suffocating, corrosive world, the prerequisites of life began to accumulate. Stirring the chemical soup of our early oceans were meteorite strikes that delivered amino acids and carbon from outer space. That warming carbonyl sulfide in the atmosphere is also likely to have condensed to form peptides and amino acids.[14] These were the essential building blocks and nutrient resources that enabled the first life on our planet to coalesce out of chaos.

Where, how, and when exactly life began in this desolate and harsh environment may never be resolved. Somewhere, perhaps in coastal shallows, deep-sea hot vents, ice crystals, or murky clays, there were stable reservoirs where conditions enabled simple organic molecules to align with each other to reproduce themselves.

Somehow, they formed self-assembling, energy-absorbing, complex structures that copied themselves and started replicating. Sometime, perhaps as long as 4.28 billion years ago, life began.[15, 16] Based on genetic evidence, the last universal common ancestor formed more than 3.8 billion years ago[17] with about 355 genes that we still recognize in organisms and ourselves today. This mystery-shrouded organism donated, among other traits, the energy-carrying proteins in our blood and the light-sensing pigments in our eyes. As humans, our genes contain these deepest of roots from the very beginning of the tree of life.

It was not long after this first magical life-forming event that the process of photosynthesis commenced—a process that was likely to have had a similar photo-synthetic machinery to that found in a modern microorganism that we call archaea today. Archaea use a pigment, identical to those in our eyes, that changes shape when it absorbs light. To create energy the pigment twists a protein, which causes acid to be pumped from one side of a membrane to the other. If an early life-form needed energy, it opened a pore in its cell membrane to let the resulting acid gradient rebalance itself, creating a flow with which to power the cellular machinery. The light sensors in our eyes use the same mechanism but, instead of storing energy when the pigment-machine is twisted by the light, it sends a signal down a nerve to our brains.

Significantly, archaea also developed the ability to produce methane, which means they were decomposing complex organic matter such as naturally occurring amino acids. These had accumulated on early Earth from the combination of atmospheric gases such as carbonyl sulfate, or from meteorites, and archaea "digested" them into simpler carbon forms before excreting them as atmosphere-warming methane.

It made sense for photosynthesis to start right away because there was little food available and assembling cells requires energy. There were no food chains; cells could not move around, so the prospect of catching and eating each other was still in the distant future. The best chance that life had was to harvest the dim energy from the sun that trickled through the murky acid clouds of early Earth. Several forms of photosynthesis co-existed at this early stage in Earth's history. Sulfur, iron, and acids were naturally abundant, and bacterial organisms developed a vast variety of chemistries that are unfamiliar to us oxygen-breathing land-dwellers today. These included the green-sulfur, purple, and other bacteria that van Niel, from the previous chapter, helped to characterize. For example, the green-sulfur bacteria thrived in the mineral-rich, acidic, dark oceans of the time. They can still be found in the depths of the Black Sea, where organic biomass decay and a lack of oxygen re-create a prehistorically dingy environment. They also flourished on volcanic ocean vents where they grow on the "black smoker," or vents of heated water.

Another life-form, found in structures in Greenland dating back 3.7 billion years, are crucial to this story. These are similar to what we now call stromatolites. Built like the rings in a tree, they comprise layers of fossilized microbial mats, which formed in the growing season, alternated with layers of sediment from being covered in mud during the winter. The microbes that formed these layers are called cyano-bacteria. This is a term you will need to remember. "Cyan" is the greenish-blue color found in your printer cartridges and, not surprisingly, these organisms used to be known as "blue-green algae." Unlike archaea, which required sunlight and organic molecules for their energy sources, cyanobacteria developed a crucial new skill by using sunlight to split water to form oxygen. There has always been a lot of water on our planet, and it is much easier to find than the complex organic molecules coming from meteorites or volcanic chemistry. The ability to split water gave cyanobacteria a powerful competitive advantage because they could thrive on fewer nutrient sources and water was plentiful.

The "Great Oxygenation Event" (and a near-death moment)

In the first billion years of their life, photosynthesizing cyanobacteria started to have a profound planet-changing impact, thanks to the waste they produced from splitting water. As these microbes grew and bacterial photosynthesis increased, they consumed carbon sources in the oceans and absorbed carbon dioxide in the atmosphere, slowly turning our world into the blue alkali oceans that we see today. The waste they produced was oxygen and, although that is now seen as an essential ingredient of life on our planet, it almost killed the other pioneering photosynthetic organisms from Earth's infancy.

Cyanobacteria accumulated along the edges of the morphing continents in their colonial, layered mud domes. As they conquered ever-more coastlines and excreted more oxygen, the oceans slowly cleared and became increasingly transparent. Light now penetrated deeper into larger areas of the ocean, more organisms could grow, and total living biomass increased.

Remember, the young oceans had been filled with brown iron minerals. But iron, when mixed with oxygen, forms insoluble compounds, and the geological record tells us that enormous bands of them were created as the gas excreted by cyanobacterial photosynthesis oxidized the iron in the oceans,[18] turning it into rust. Around 2.8 billion years ago, the first of a series of very large bands of iron deposition attest to large-scale photosynthetic activity releasing ever-increasing amounts of waste oxygen.

The vast amount of iron in the oceans had acted as an enormous chemical buffer that absorbed this waste oxygen. However, when the oceans ran out of iron, the waste oxygen began to accumulate in the atmosphere and, in doing so, almost wiped out most of life on Earth. This atmospheric accumulation of oxygen 2.4 billion years ago has been named the "Great Oxygenation Event." Over a period of four hundred million years, between 1 and 2 percent of the atmosphere filled with oxygen. This may not sound much when compared to the 20 percent of oxygen that we breathe today, but to most of the early life forms fighting for existence at the time, it was a significant amount, and a pollutant.

It may seem paradoxical that oxygen could have been a pollutant. Yet, oxygen is also a highly reactive molecule that oxidizes or "rusts" susceptible materials such as cells or iron. Part of the reason why we care about the antioxidant quality of our foods is that oxygen is toxic to cells, and there are enzymatic and chemical safeguards against reactive oxygen damage that have evolved in response. These protective mechanisms cost cells a lot of energy. For example, in photosynthetic cells, the biochemical process of eliminating oxygen also creates toxic ammonia, which then in turn needs to be cleaned up. These oxygen management activities reduce the overall efficiency of cells.

Another, even more pernicious, impact of oxygen that still plagues us today is that it slows photosynthesis via a mechanism called photorespiration, in which it competes with the carbon dioxide for the carbon-absorbing machinery. This photosynthesis-blocking activity of oxygen reduced the capability of other microorganisms such as archaea—which had evolved in an acidic world without oxygen—to grow.

An epic tug-of-war began between two models of photosynthesis. On one side were the archaea, which were so well adapted to the world as it had been. On the other were the cyanobacteria, which were polluting the atmosphere with all that oxygen and gradually transforming the world.

As oxygen levels rose, the archaea found it harder to photosynthesize and produce their planet-warming methane. At the same time, the growth of cyanobacteria meant that more carbon dioxide was being consumed and bound in stromatolite-type microbial mud domes or, in other words, removed from the atmosphere. Because of the decline of both the major atmosphere-warming gases—carbon dioxide and methane—the atmosphere no longer trapped as much heat as it had done before. To make matters worse, oxygen is highly reactive in the upper atmosphere where, together with water vapor, it breaks down methane.[19] This was at a time when overall volcanic activity was calming down as well, further reducing the amount of methane in the atmosphere.

During the Huronian glaciation period 2.4 to 2.1 billion years ago, the decline of both of the major atmosphere-warming gases meant Earth was cooling. By this time, the other globe-warming gas, carbonyl sulfide, which had helped form a warming blanket over our young planet, had also been consumed by the sulfur-processing bacteria. As a consequence, our entire planet froze, and Antarctic temperatures reached the equator. For three hundred million years, Earth was covered with glaciers, and this monster ice age became a vicious cycle as white snow reflected the weak sunlight. The life that was left on the planet was concentrated into small pools of liquid water trapped under the ice—maybe it clung to deep-sea volcanic vents or froze into spores stored in the ice sheets.

The great irony of photosynthesis, the process that turns light into life, was that it almost killed life on our planet even before complex organisms could begin to evolve.

Planetary rebound

We don't know exactly why the Earth began to thaw out again, but it is likely that it was because atmospheric carbon dioxide slowly began to accumulate. The carbon build-up could have been due to volcanic activity and decay of the previously formed biomass. It is also probable that volcanic dust darkened the ice and accelerated snow melt. Some evidence suggests that continental landmasses shifted to the poles, thereby opening oceans for light absorption and reducing the amount of snow reflecting the light.

Those same photosynthetic cyanobacteria that had almost wiped out life on Earth may also have redeemed themselves by coming to the rescue[20] in the form of newly adapted snow-dwelling photosynthesizing algae. The snow algae darkened and sculpted the snow, absorbing more sunlight and helping the planet thaw. You can still see carpets of these snow algae in the far north or on mountain slopes where they dramatically color the snow in supernatural-looking pink—accelerating the snow melt of Greenland and Antarctica today.

The great freezing of our planet had another consequence, too. The shifting ice sheets scouring the planet resulted in cracking and tilting landmasses. Mountain ranges and enormous sedimentary basins took shape where lakes, inland seas, and all kinds of other environments for biological activity emerged.

In these new landscapes and oceans, the first complex organisms with multiple internal cellular compartments formed. They permanently hosted bacterial cells within their structure, resulting in layers of membranes and cells becoming integrated to the point where they joined up to make entirely new organisms. By 1.56 billion years ago, evidence of the first multicellular structures with shapes reminiscent of modern seaweeds appeared and were fossilized.[21]

As the ice sheets retreated, the super-continent Rodinia, which more or less filled the space we know as the Atlantic, split and drifted apart to form our more familiar continents, with new climates offering ever-greater environmental possibilities. It was during this period that sexual reproduction began, with the consequent ability to adapt, combine, and create greater biological diversity.

But then, between 720 and 635 million years ago, Earth froze over again as these newest oceanic algae depleted atmospheric carbon dioxide once more and precipitated another long ice age. For a second time, photosynthetic and other organisms found shelter or protected niches or developed other ways to acclimatize and survive. Genetically, what was there before seems to have continued unabated and, as the ice melted between 635 and 541 million years ago, new and even more complex life-forms emerged.

The first animals were probably something very similar to modern comb jellies—sometimes known as sea gooseberries—little balloon creatures that fan microscopic organisms into simple mouths. Alongside them, another strange new life-form evolved as the first plants derived from green algae oozed out of the ocean and populated the land. For the first four billion years of Earth's existence, land had been almost entirely free of life. But now sea algae began to leave the water and colonize the coast. Algae entered into a symbiotic relationship with fungi to form the simplest of simple mosses and lichens[22] that could live on land. Even today, some of these early pioneers still live in the Antarctic's dry valleys. Not only did they cultivate new space, they also brought with them the ability to "fix" nitrogen—the process by which huge quantities of the gas found in our atmosphere is turned into protein and cellular components. Simple mosses soon evolved and began excreting organic acids, which accelerated the weathering of rocks. As Earth's crust shifted and migrated, these new organisms populated every crack and crevice they could find. Even unyielding material like granite was rapidly degraded by this biological assault, which broke down enormous amounts of land-locked minerals and nutrients.

These minerals then washed into the ocean and, in time, fed such large algal blooms that atmospheric carbon dioxide plummeted for a third massive ice age 545 million years ago. Once again, the stunning success of photosynthesis had put life back in the freezer.[23] And this time, most of the latest innovations in life—including the new complex organisms—were driven into extinction. For the first time, large branches of the "tree of life" withered and died.

And yet some new photosynthetic organisms survived to reclaim and help thaw the planet. They even developed new skills to populate the remaining living spaces on the previously unpopulated land, or another 30 percent of Earth's surface.

I used to imagine that it would be impossible for anything to live through an ice age lasting hundreds of millions of years. That was until I worked under the ice through two Antarctic winters and springs. In the early 1990s, as a graduate student, I had gone to the Antarctic to help measure the UV-radiation baseline before the Ozone Hole formed. What seemed like a life-threatening, deep-frozen ice desert turns out to conceal a lush ecosystem beneath the surface. Marine photosynthesizers thrive not only in the seasonal marginal ice zone but also in the multi-year pack-ice of the North and South Pole. These are particularly productive environments because the underside of the ice surface, to which the organisms are attached, is transparent to most light and prevents them from sinking to the dark depths. It is also relatively stable and permanently washed with a rich supply of nutrients from the ocean. Among the animals that migrate there to feed on these abundant algal gardens are whales, penguins, vast swarms of krill, seals, seabirds, walruses, as well as crustaceans, mollusks, and countless fish. While in Antarctica, I also chased archaea. At the time, the conventional wisdom was that they were "extremophiles" because we knew they were found in sulfurous hot springs such as the steaming geysers of Yellowstone National Park or in hypersaline lakes. Instead, the Antarctic water demonstrated that they also live happily under the ice, in great abundance, and populate all layers of the ocean.

Given all the problems with periodic ice ages that oxygen pollution from photosynthesizing cyanobacteria was causing, life on Earth might have evolved differently. Although a world without oxygen would not have sustained large, warm-blooded animals that burn a lot of metabolic energy like us, life on the planet would still have functioned perfectly well at a slower pace. Fortunately, oxygen also has some major advantages. It enables fungi, bacteria, animals, and plants to convert food into energy at a much more efficient rate than can be achieved without it. Fermentation— where microorganisms break down complex sugars into simpler compounds such as alcohol—is the next best metabolic strategy, releasing two energy-carrying molecules from one simple sugar molecule, for example glucose. In comparison, metabolism with oxygen is nineteen times more efficient and breakdown of the same glucose produces no fewer than thirty-eight cellular energy-carrying molecules. Ecosystems grew everywhere because the overall benefit of oxygen in the atmosphere outweighed the additional cost of detoxifying it. It meant more organisms grinding up more rocks, making more nutrients available for yet more organisms, creating more biomass and, ultimately, spawning a greater variety of living creatures in a globe-spanning virtuous circle.

The other essential ingredients for life

To feed a global population of plants, animals, microbes, and people requires huge amounts of mineral resources—the natural fertilizers that drive global ecosystems. Plants do not, as was once believed, eat dirt. But they do need some dirt in which to grow. And, importantly, they also need to move it.

The planet's surface wears down slowly as it is eroded by wind, ice, rain, waves, and gravity. Water penetrates rocks and, when it freezes, can crack even massive ones open. Rain absorbs carbon dioxide, is slightly acidic, and eventually dissolves many minerals. Physical forces like frost, thermal stress, wind, or waves also break down hard surfaces over long periods of time. Biological weathering dramatically enhances this process. Roots physically crack rocks. They also excrete organic acids that dissolve difficult-to-access minerals. Photosynthesis-generated oxygen breaks down minerals and makes them available for dissolution. These are transported by rivers to the oceans, where silt and sediment movement creates enormous deltas on the edges of continents.

Estimates for the cumulative total rate of these different forms of weathering range from 0.01 to 0.5 mm per year for every land surface on Earth. For silicate minerals—such as limestone or granite—this amounts to about 3.7 billion tons of dissolved rock that is flushed via our rivers to our oceans every year. Now add silt, dust and many other mineral movements, and approximately 2 mm of Earth's land surface is washed into the oceans each year. That might not sound like much, but it means that every decade approximately 2 cm—the best part of an inch—of its surface has been washed away. It is these dirt movements that release the minerals necessary to feed every ecosystem in the world.

Not surprisingly, the largest contributors to mineral movement are the Himalayas and the Andes. Large, steep mountain ranges erode faster because they are exposed to greater temperature fluctuations and gravity moves the materials the farthest. The formation of the Himalayas and the Andes, which started fifty million years ago, directly correlates with the formation of Antarctic ice. The extra minerals that washed down the deep slopes of those mountains entered Earth's systems. They not only absorbed carbon themselves but provided the nutrients to grow more photosynthesizers that, in turn, absorbed even more atmospheric carbon dioxide. The result was a cooling of the southern hemisphere and the forming of the Antarctic ice sheet, which has lasted for thirty-five million years.

Some geoengineers believe that a method to improve our environment would be to deposit a continent's worth of powdered limestone in the ocean. The idea is that if we can make more calcium and magnesium available, they will bind excess carbon dioxide from the atmosphere. The concept is called "enhanced weathering" or "ocean alkalinity enhancement."

I am not so sure that grinding up huge areas of useful land—for example southern Australia—to dump into the ocean is such an elegant solution. It might be better to get land plants to do the work for us. We could help plants to grow in previously uncultivated places—and they could settle on everything from newly formed volcanic rock to the smooth sides of a modern city skyscraper—where they would not only absorb carbon dioxide themselves but also grind up the rock to create larger nutrient cycles when the rain returned these minerals to the ocean.

Balancing this virtuous cycle is tricky. When too

many nutrients wash into the ocean via our rivers, we sometimes get disruptive algal blooms. For example, the Amazon River now carries agricultural fertilizer that feeds vast rafts of kelp that ultimately wash up on Caribbean beaches. But it can be done in a controlled way. For instance, deltas are among the most productive land areas on Earth. When river water mixes with the salty seas, most of the dissolved minerals precipitate to form new and highly fertile land. Think of the Amazon, Nile, or any other delta. If we get better at managing silty river run-offs, we can deliberately form new land to feed the world and take carbon dioxide out of the atmosphere. Another natural trick that can help arid areas to turn into more productive ones is to make better use of oceanic algae. Purple bacteria create a garlicky, slightly fetid odor when they break down organic sulfur compounds in the ocean. This smell turns out to be a molecule that helps water vapor condense in clouds. The more these organisms stink up the ocean, the more clouds there will be drifting on to land, the more rain to help grow plants, and the more minerals from the soils carried back into the ocean.

As for the nutrients created by this cycle, the most important one is nitrogen. This is essential for all living organisms because it is the key ingredient in proteins and many other biological molecules. Historically, nitrate that washed from the land into the sea was then absorbed by the ocean's photosynthesizers. The algae-based nitrogen then moved up the food chain until it ended up in the droppings being generously deposited by marine birds on land. These droppings— or guano—closed much of the nitrogen cycle, especially when it washed back into the ocean. The annual nitrogen cycle in the form of dissolved aquatic and marine minerals is approximately 2.8 billion tons. In recent centuries, busy humans have used up the guano and an enormous amount of energy to make synthetic fertilizer and grow bigger crops. It is estimated that half the nitrogen in each of us human beings comes from this new man-made source.[24]

Another essential nutrient is phosphorus, which is found in DNA and cellular information molecules. However, phosphate deposits are less common and are derived from ancient marine sediments. The largest phosphate mines are found in Morocco and contain enormous numbers of fossilized sharks' teeth, revealing how they formed from marine organisms piling up over millions of years. These days there are only around twenty million tons of phosphate minerals moving through the system per year, and we are mining a very limited historic source. A shortage of phosphate is especially challenging for land plants, which matters a lot in a world where we need to produce even more food. However, large amounts of phosphorous are currently wasted because we allow it to flush into our rivers and oceans via our wastewater and agricultural run-off. Phosphate is the predominant cause of eutrophication—an excess of nutrients—that can cause unwanted algal blooms. The solution is for us to use wastewater technologies that capture the phosphate before it is thrown away, preventing a harmful excess of this critical nutrient where we do not want it and providing a valuable source of fertilizer where we need it.

A third crucial nutrient is iron. I have a friend on a desert island in the Caribbean who is obsessed with creating a better garden in his very sandy soil. He has created *terra preta*, or black soil, from kitchen waste, and is very proud of the exuberant abundance of his flowers. There is, however, another reason why his garden is so lush: His soil also contains sand that has blown in from the African Sahara on the other side of the Atlantic Ocean, which contains traces of iron. A shortage of iron lowers overall photosynthetic activity, especially in the ocean. Sandstorms from deserts around the world are a source of iron essential to the growth of marine ecosystems.

These days, of course, the essential nutrient for photosynthesizing organisms is carbon. On land, biomass in the form of plants and wood can be eaten or burned. Still more is buried in the soil as part of a natural, steady, and slow process of building up carbon reserves. We usually know these carbon deposits as coal, oil, natural gas, or peat, and they are even found in permafrost. The slow accumulation of organic remnants in these relatively stable stores ensures long-term removal of carbon from the atmosphere. It is vital for the sustenance of life. Indeed, the total amount of carbon that is "fixed"—or absorbed into biological organisms—is a stunning 123 billion tons per year. This is *the* big cycle and involves approximately a trillion tons of live organisms freshly growing on our planet using photosynthesis every year, and then moving enormous resources across the land surface of our Earth and through our oceans. A trillion tons of live biology can, quite literally, move mountains.

The ocean is once again critical to storing carbon because seawater stores 93 percent of available carbon dioxide. Oceanic carbon is found in three major forms: free carbon dioxide as a gas; bicarbonate, which makes the bubbles you find in sparkling mineral water; and soluble carbonate minerals from the dissolved rocks, which make the seawater salty. As photosynthetic organisms grow in the ocean to form biomass, they accumulate carbon in their cells and structures. The result is carbon-depleted seawater that is then available to absorb fresh atmospheric carbon dioxide. At this point, let me introduce you to two forms of algal seashells called coccolithophorids and diatoms that perform this vital task really well. I will discuss them in more detail later on, but they are particularly important in this carbon cycle. Coccolithophorids include cells that are covered with little overlapping shield-like plates. These plates, made of calcium carbonate, form beautifully intricate defensive cell walls. There are deposits all over the world from marine sediments of this prolific and calcium-absorbing algae, including the famous White Cliffs of Dover. Diatoms, by contrast, have glass-like shells made of sodium silicate for their protection. They are heavy and, at the end of their life cycle, they fall to the seabed, where sediments build up. Together with corals, shell-forming mollusks, and other marine microorganisms, these deposits are eventually converted into limestone, marble, and other forms of stone.

All biological organisms reflect the nutrients they can access in their environment. Living things are all mostly made of a mixture of water and the following ingredients: carbon, which is found in almost all organic compounds; nitrogen, which is necessary for our proteins, muscles, and genetics; phosphorous, which is found in our DNA and cellular energy carriers; as well as traces of elements such as iron. Unsurprisingly, there are deviations here and there. Trees contain disproportionately more carbon in the form of the cellulose or wood that they store in their trunks. Mammals like us have heavy bones that are rich in calcium. Land animals originated in the ocean and carry it with them. Even our human blood has the same salinity as that of the ocean at the time when our genetic forebears first colonized the land. In the oceans, there is a useful measure known as the Redfield Ratio, named after the man who discovered it, Alfred C. Redfield of the Woods Hole Oceanographic Institution. The ratio indicates that marine organisms have the same elemental combination in terms of carbon, nitrogen, and phosphorus as the seawater in which they live. Marine organisms are physically made of the materials they can access in the ocean. Ultimately, these are the elements photosynthetic organisms wrest free from the planet—and they are also the ones that they move around the planet to the ultimate glory of life. In essence, the old saying that "we are what we eat" is true.

CHAPTER 3

THE PHOTOSYNTHESIZERS

It is unsurprising that as land-dwellers humans tend to assume that photosynthesis is about flowers, vegetables, mosses, trees, grasses, bushes, and weeds. These photosynthesizers represent an enormous variety of organisms but, as I mentioned earlier, not everything that captures and processes sunlight is a familiar green plant. The magical process of converting light into life happens everywhere, from the floor of the deepest ocean to the top of the highest mountain.

As with all biological systems, there are evolutionary twists and turns as multiple photosynthetic systems have been absorbed, incorporated, modified, or combined. This story is about a community of species that collaborate, compete, or prey on each other. Photosynthetic organisms have reemerged, split, and recombined frequently. There are countless mergers, hostile takeovers, and unexpected partnerships.

In recent decades, scientists like Sallie (Penny) Chisholm or Ed DeLong, both at the Massachusetts Institute of Technology, have discovered that bacterial photosynthesis can be found in vastly different forms almost anywhere on our planet—in acid mining tailings and alkali lakes, in salt- and freshwater, in hot springs and Antarctic cliffs. Photosynthesis thrives in the presence of oxygen or in its absence, and with all kinds of carbon sources other than carbon dioxide. Indeed, these days the simple water-and-carbon-dioxide-produce-sugar equation no longer suffices to explain the range of photosynthesis. Instead, an enormous variety of organisms have incorporated the skills to convert radiant light energy into chemical energy. To make things more complicated, we are still finding new photosynthetic processes that we do not yet fully understand. Some have only been discovered since we gained a satellite-based global perspective, and we have yet to work out the mechanisms of many natural ecosystems. Where we thought we had a simple bell, we recently discovered that what we actually have is an orchestra with myriad possible tonal and chord combinations.

Hidden in this bewildering variety of organisms are endless opportunities to utilize newly discovered biochemical processes, develop alternative production methods, and, yes, fix our environment, too. The last century was all about chemistry. The one before that was all about physics. In this century, sophisticated biological processes could yet change every manufacturing and production industry so that we become not only more effective at looking after ourselves but also able to clean up the mess we have made of our planet.

Our new friends, bacterial photosynthesizers

Ralph Lewin, an Anglo-American geneticist, was teaching marine microbiology at the Scripps Institute in California in 1975 when he discovered a cyanobacterium called *Prochloron*,[25] of which there are phenomenal numbers in the ocean. Because they are minuscule, these photosynthetic bacteria are particularly good at absorbing very small traces of nutrients left over by larger organisms. They thrive even in the nutrient-depleted surface waters of the open ocean where other organisms can no longer grow. These tiny cyanobacteria are just one of a number of lifeforms that are exciting biochemical engineers as they seek new ways to produce crops, or just clean up our environment.

In the early 1980s, less than a decade after Lewin's discovery, a kind of brownish-green *Heliobacteria* was discovered.[26] This type of bacteria cannot live in the presence of oxygen and is found in the mud of rice fields, Siberian soda lakes, and hot springs. The bacteria are very effective nitrogen fixers, which means that they can turn atmospheric and inert gaseous nitrogen into organically available compounds. Instead of using carbon dioxide, they can thrive on a large variety of organic carbon sources. They also help to fertilize rice plants by fixing nitrogen and thereby directly contribute to improving the soils that feed a large part of the world's population.

Acid-tolerant soil bacteria were not discovered until the 1990s.[27] These acidobacteria were found in acidic iron-rich mine drainage pits where the conditions resemble the conditions of the dark oceans of our early Earth. They are found in soils, clays, and aquatic environments all over the world. Their ability to digest iron-containing compounds means they make an important contribution to soil fertility and cleaning up metal contamination.

In the early 2000s, filamentous strands of a form of bacteria called *Chloroflexus* were identified. These organisms thrive in alkaline hydrothermal vents at temperatures of up to 160°F, which would scald human skin. They contain fundamentally different light-processing machinery and have a different carbon-dioxide-fixing method, too, making them attractive for carbon sequestration schemes.[28] Interestingly, these creatures can also break down environmental toxins.

And then as recently as 2007, another group was discovered that represent up to 10 percent of the bacterial population in oceanic surface waters.[29] They live in oxygen-rich environments but, somehow, also perform photosynthesis without oxygen as they mine organic carbon from the ocean. Given their abundance, it is likely they have a significant impact on converting complex oceanic carbon into the biomass that feeds fish and finds its way to the bottom of the sea. Even ancient organisms that we have known about for a long time are showing they can surprise us. It turns out that archaea can live on hydrogen gas, ammonia, or even metal ions. There is a branch of archaea that lives in salt lakes containing between 10 percent and 37 percent salt. In comparison, the pickles in your refrigerator are preserved in 1.2 percent salt. These extra-tough archaea have a dark salmon-red pigment that is used as a highly effective sun block.

Some of the ancient bacteria from the beginning of Earth's history were found in 2004 at a depth of a mile and a half (2,500 m), living on the faint infrared glow of a volcanic black smoker.[30] These purple bacteria thrive when there is no oxygen but a consistent supply of sulfide from volcanic or decaying organic matter. This makes them particularly useful for breaking down complex and toxic organic compounds in soil.

As for our old friends, cyanobacteria—formerly known as blue-green algae—you may actually have eaten it in the form of spirulina, a dietary supplement found in lots of health shops and used as a food colorant for syrups, sweets, and beverages. These bacteria have medicinal properties as nerve and liver protectants and are used in anti-cancer anti-inflammatory drugs. And because they have mastered how to fix atmospheric nitrogen, they can also be converted into organic fertilizer.

Working together

In recent years, scientists have been discovering that cooperation and collaboration are at the heart of life itself. And crucial to our developing understanding of photosynthesis is the idea that the bacterial photosynthesizers live inside other cells for the benefit not only of themselves but also their hosts.

For most of the last century, this was a heretical scientific opinion. The orthodox view was that nature and evolution were driven by competition.

The first pioneer to challenge conventional thinking in this area was Andreas Franz Wilhelm Schimper, an accomplished explorer who loved brightly colored organisms. Before he died of malaria in 1901, Schimper spent a lot of time carefully counting the layers of membranes around the outside of chloroplasts. This led him to hypothesize that chloroplasts are actually photosynthetic cells in a symbiotic relationship with more complex organisms.[31] To him these "chlorophyll-grains" provided a "pigmentation" service to the otherwise bland host organisms.

In 1910, Konstantin Mereschkowski turned Schimper's idea into a theory, which he called Symbiogenesis. He argued that multiple bacteria and amoebae jointly form cellular organelles—the nucleus and chloroplasts—to make the "higher" organisms we recognize in plants.[32] This revolutionary theory went largely unnoticed at the time because Mereschkowski, although a good biologist, was a deeply flawed human being who also proposed eugenic breeding for a society that would cull its members at the age of thirty-five. Fleeing prosecution for the rape of dozens of girls, he finally committed suicide in accordance with his twisted dystopian nightmares.

On a happier note, Ivan Wallin was an American-Swedish biologist known for his extravagant herring and lutefisk parties, and adored by students for his unconventional style of lecturing. In a series of papers from 1922, he set out theories that mitochondria—the energy-producing organelles within cells—are, in reality, "bacteria or bacteria-like organisms that accepted the leisure of a symbiotic partnership" with higher organisms.[33, 34]

Such ideas were still regarded as dangerously unorthodox until the late 1970s, when Lynn Margulis, an American evolutionary theorist, overcame fierce resistance to prove definitively the bacterial origin of chloroplasts and other cellular components. Margulis's tenacity and persistence in the face of being rejected no less than fifteen times by scientific journals is still inspiring. She helped collate many different strands of research and then personally supervised the translation of previously unknown theorists including the despicable Mereschkowski. She championed the idea that symbiosis is the fundamental driver of evolution,[35] with cells working together at all levels up to the formation of entire ecosystems. As Margulis put it: "Life did not take over the globe by combat, but by networking."

This networking—known as endosymbiosis—is often portrayed as a larger cell engulfing a smaller one. You could also see it as a smaller organism moving in with a larger host. These relationships can be exploitative or casual but generally they are cooperative and mutually beneficial. Organisms gain resources, protection from predation, and other advantages that represent a win-win for all partners. Such organisms show that by working together, their chance of survival is better than if they lived on their own.

Think, for a moment, about the unremarkable stunted tree, *Camellia sinensis*. Its collaboration with one particular animal has taken it from its rain-soaked, tropical East Asian origins to the highest mountain meadows across the world. By providing human beings with delicately flavored caffeine, *Camellia sinensis*—the tea plant—has increased its biological range, growth rate, and survival, dominating many more mountain slopes than it could ever have done by itself.[36] This is not a conscious collaboration. It is one that is selected by contextual environmental forces where both organisms benefit from each other.

Creating ecosystems

The first act of cellular cooperation happened very early in Earth's history and, since then, countless microorganisms have joined complex symbiotic linkages to produce the spectrum of modern organisms we know today.

They are like a tangle of telephone wiring but, by sorting through the different color pigments and pulling back layers of membranes around the chloroplasts, it is possible to work out where connections took place and begin to categorize them. For the purposes of this book and the sake of simplicity, we can place these photosynthesizers into three groups.

The green branch includes green plants but is not always green. For instance, this group includes the bright-pink cells that dramatically color salt lakes and protect organisms from the sun. Himalayan, or pink, salt was stained by these cells as it dried into salt deposits. The point is not to take these colors too literally. These organisms have learned to be green in a vast range of environments, except when sometimes it is better to be pink. It is, of course, a branch of life that provides us with the food we grow, about half of the pharmaceuticals we take, as well as the forests, swamps, fields, hedges, and gardens that we call home. It is also the largest store of carbon biomass on Earth's surface, providing the balance we need on our coastlines, forests, and overused land.

The red branch is familiar to us as seaweed or the agar that jellifies our puddings and custards, but it can be dated back at least 1.2 billion years through fossils of kelp-like structures, which probably makes this kind of seaweed the oldest multicellular photosynthesizer on Earth. It is a branch that represents more than seven thousand freshwater and marine species, including the nori in which we wrap sushi. Compounds from red organisms turn out to contain many potential pharmaceutical benefits for humans. As a feed supplement, they reduce methane emissions from cattle, and farmers across the world upgrade their soils by adding seaweed extracts. The red branch includes the coccolithophorids we met earlier in this book, which make the little shells that eventually form chalk deposits, including those at the White Cliffs of Dover. Vast layers of these algal sediments have periodically cracked continental plates with the weight of their shells. Blooms of them bind huge amounts of atmospheric carbon dioxide and, when they die, their shells sink to the bottom of the ocean as relatively inert and stable chalk.

There is another enormous group from this red branch of photosynthesis composed of at least fifteen thousand species. All of them have a symbiotic relationship with organisms inside them. They include not only the large brown clumps of brown kelp we know from beaches around the world but also bloom-forming organisms that are responsible for feeding most of the fish and animals in our seas. And they make almost half of the oxygen we breathe. Some of these algae are the diatoms that form unbelievably intricate silicate cell walls, about which nanomanufacturing material scientists can only fantasize. These inorganic glass cell walls form fine sediments and sands. Such "diatomaceous earths" formed the basis of Alfred Nobel's wealth, with their ability to stabilize explosively volatile nitroglycerine so that it can be shaped into safe dynamite sticks. Today these glass shells are being evaluated for more peaceful purposes such as cancer-drugs and antibody delivery tools.

Many of the most abundant organisms in the world are so plentiful precisely because they are photosynthetic. Being able to feed themselves makes them thrive, and all of this is due to the collaboration of a photosynthetic bacterium inside another non-photosynthetic cell. Without the ability to feed themselves, the host organisms can become nasty predators. The host organism, which in photosynthetic collaborations makes much of the oxygen we breathe and feeds the oceans, will, left to its own devices without its photo-synthetic partner, destroy our crops and starve our species. Red algae without their partners include water molds such as *Phytophthora*, also known as the deadly potato blight, which was responsible for the nineteenth-century Irish Famine.

When non-photosynthetic, collaborative organisms—or host cells—lose their photosynthetic partner, they are able to squeeze between other cells to act as gliding parasites. These parasites can be the source of terrible diseases such as malaria and toxoplasmosis.

These red-and-green pigmented collaborations are among the most powerful branches in the tree of life. The enormous expansion of life-forms and the conquest of land all stem from the moment the first photosynthetic bacteria were absorbed into a non-photosynthetic, non-bacterial host organism, which then began working together in endosymbiosis. But evolution did not stop there. Instead, the collaborative organisms took residence inside other hosts, a bit like the nested dolls of Russian matryoshka. This is the third group of collaborative photosynthesizing organisms, a vine that twists itself around the core green and red branches.

Unfortunately, this lineage usually comes to our attention as the so-called harmful algal blooms or "toxic red tides" described breathlessly by the media as beaches are shut and marine life killed. This same group makes much of the bioluminescence that lights up the surface of oceans and represents the height of complexity in single-celled organisms. They not only have the ability to produce light but also form complex "eye spots" that allow them to follow the sun and, with their two strong flagella, swim in a purposeful direction. Although they are neither the fastest-growing form of algae nor the best at photosynthesizing, they compete effectively because, unlike other algae, they can swim through different temperature layers in water to absorb deep-lying nutrients at night and then rise back to the surface in the morning to photosynthesize in the light. In so doing, they increase the overall carrying capacity of the sunlit surface waters to produce more biomass than other algae are able to.

The vast majority of these organisms are harmless, so much so that fish like sardines time the spawning of their eggs to coincide with their reliable blooms. However, a handful of them use other tricks to survive. Some can poison the water for their predators or competitors with chemicals so toxic that the faintest trace of their vapors causes closure of beaches and in high doses can be fatal to humans. *Pfiesteria*, or the "fish-killer," is a particularly versatile creature that boasts up to twenty-four different life stages.[37] When there are inorganic nutrients about, their cells photosynthesize. When neither sunlight nor nutrients are plentiful, Pfiesteria transforms itself into an amoeba and attaches itself to a fish, which it will proceed to poison and digest. As the fish begins to leak vital juices, the amoeba enters a sexual phase. The daughter cells feed off the fish, usually until it dies, before spawning into the water again, where they will resume conventional photosynthetic activity as if this flesh-crazed orgy had never happened.

This versatile third group of organisms have an extraordinary ability to survive our complex and challenging world. They have even been seen hitching a ride on the outside of the International Space Station to which they stuck electrostatically after dust from evaporated seawater containing them drifted out of Earth's atmosphere into low orbit.[38]

It's important to underline that there are many different types of photosynthesizers living in very different environments, doing so much more than their essential work of absorbing carbon dioxide from the air and growing our food. They can break down residual carbon in seawater or export it to the ocean bottom. They can purify contaminated soils and clear up industrial waste. They can fertilize rice fields and produce all manner of valuable products other than food.

As a result, there is a whole range of further and largely unexplored chemistries available for carbon storage, nitrogen fixation, or to feed complex food chains. We can tap into these alternate processes to develop sustainable industries or repair our environment. Like cities, ecosystems are the product of many specialists, from many different calls of life, working for a common objective. The bigger and more creative the city, the richer, healthier, and more productive it is. In the biological cities of ecosystems, collaboration ranges from simple partnerships between two organisms to entire networks of competitors, predators, and recyclers working together.

Long before any organism colonized land and long before the word "economy," this process was enabling more efficient utilization of resources, new forms of constructive competition, and extraordinary levels of growth. Today, symbiosis and sharing of genetic information is creating new products, new services, and new business models. In nature, as in business, there are always some predators, thieves, and freeriders. Yet the network almost always finds a way to limit their impact.

Not so tiny photosynthesizers

It is easy to believe that in our modern world the dominant agent for change is human. This book has already referred to people "discovering" new and unexpected organisms rather than them discovering us. At the same time, our news is filled with grim reports of environmental catastrophe—pollution, disasters, disease, and famine—caused by humans. But, for most of Earth's history, photosynthesis had been the most powerful force for transformation. And, even since the advent of the human era, photosynthesis has not stopped changing our world. It is my belief that it can yet help us solve the problems we have caused. Could we see creeping kelp invade land, with cities coated in floppy fronds as this light-hungry plant seeks brighter heights? Or could snow algae evolve to carpet Greenland in pink and red meadows, fertilized by the newly carbon-dioxide-enriched air? Photosynthesis has broken species boundaries before. Could it fuel animals so that they could sunbathe instead of preying on other creatures?

Corals, snails, aphids, insects, and salamanders

These questions may seem far-fetched until you meet some of the larger organisms that are photosynthetic, including some of those still learning the trade—because evolution is far from finished. We have always been taught that there is a duality in natural systems, where primary producers are consumed by predators and recyclers. However, we are now finding that even animals have succumbed to the charms of photosynthesis. The reason this is obvious: Photosynthesis, which provides so much clean and free energy, makes sense for everyone.

Corals have been on the planet for more than five hundred million years, longer than almost any other living animal. They are ancient marine predators that, along with the similarly long-lived anemones and jellyfish, distinguish themselves by harpooning prey with microscopic stingers. But that is not how they get most of their energy. Corals mineralize their soft bodies into semi-hard, tree-like structures, forming calcium carbonate skeletons that provide an ideal protective habitat for symbiotic photosynthesizers. These feed on the carbon dioxide and nutrients excreted by their coral hosts. In return, the corals get as much as 95 percent of their energy in the form of sugars made by their symbiotic partners. Corals live in coastal seawater that has already been depleted of nutrients by other organisms. The photosynthetic symbionts in corals may not be the fastest-growing organisms. Yet their collaboration results in a sun-powered, turbo-charged system in which every living thing benefits. Even the zooplankton—microscopic prey that is eaten by the corals to top up their diet—thrives in an ecosystem with lots of hiding places, complex food chains, and stable, if low, nutrient supply.

Figure 2: Corals are complex colonial organisms that incorporate photosynthetic dinoflagellates (zooxanthellae) around a flexibly calcified structure. In combination, corals and their photosynthetic symbionts create some of the most productive environments on the planet. In terms of photosynthesis, the coral reefs built in sunlight-drenched and shallow coastline waters are the single most productive habitat on the planet. They lead the carbon fixing league tables, converting 2,500 grams per square meter per year—making them 3.6 times more productive than boreal or mid-latitude forests.

The benefits to humans are endless. Coral reefs protect our coastal cities from storms and tidal waves, support a vast variety of nutritious fish, and can create jobs in sustainable tourism. Most importantly of all, in a world of ocean acidification and atmospheric pollution, they are the most effective carbon absorbers we have. Instead of nurturing these coral reefs, however, humans are actively destroying them, with careless unsustainable tourism, industrial development, and the mismanagement of predatory organisms. Worst of all, thermal stress from our warming oceans is causing corals to bleach—a panic response to excess heat in which corals expel their photosynthetic symbionts—which means they slowly starve to death.

Snails are animals that have much more complex nervous and digestive systems than corals or jellyfish. They have also mastered the trick of incorporating photosynthetic organisms. Some sea slugs, a type of snail, employ a special version of "kleptoplasty"—a word derived from the Greek for "thief"—in which they eat green algae. They extract the chloroplasts but, rather than digest them, they then maintain this stolen photosynthesizing machinery in their guts. Unlike corals, they do not appear to derive a great deal of energy from their captured chloroplast. Instead, one of the advantages for the slugs appears to be that the chloroplasts turn them bright green and provide them with camouflage protection from predators in the weedy salt marshes and tidal flats where they live. However, the story does not end there. The chloroplasts persist inside the snails for up to nine months, longer than they would have lived in the algae themselves. The chloroplasts are metabolically active and continue to self-repair. This should give us pause for thought because when chloroplasts, which used to be independent cells, enter a symbiotic relationship, they relegate much of their genetic competency to their hosts, so that the two can cooperate effectively. This means that the host regulates the maintenance of the chloroplasts. In the absence of the algal host, which has been nibbled off by the snail, how can the chloroplast continue to thrive for many months in the animal gut cells?

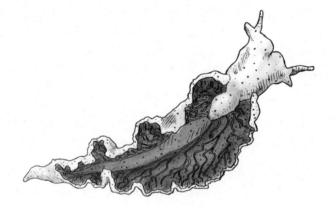

Figure 3: The leaf-shaped, leaf-sized sea slug *Elysia chlorotica* is bright green from the green algal chloroplasts it retains in its body for up to nine months. This provides green camouflage and metabolic support. At the same time, the stolen or extracted chloroplasts are supported and fed by the snail's cells. This means that the snail has adopted the genetic machinery to support its photosynthetic endosymbionts as a plant would.

The surprising answer is that the snails have absorbed and adopted the necessary chloroplast-enabling genes from the algae into their own genome. In other words, these animals are learning how to become plants.

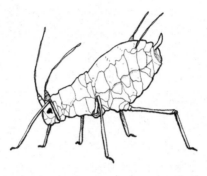

Figure 4: The garishly bright-red or bright-orange pea aphid *Acyrthosiphon pisum* synthesizes lycopene (just like tomatoes). Instead of absorbing the pigments, it has incorporated the pigment genes and synthesizes the pigments itself. This enables it to absorb light to power its mitochondria (in intracellular energy organelles).

Our journey continues with the aphid. These little insects suck the juices from plants and excrete tiny droplets of sugary water that stick to everything. Personally, they upset me greatly when they sap the plants in my flowerpot, devour young rosebuds, or choose to infest a tree under which I have parked my car. My son collects ladybugs because they eat aphids. At times, I have to confess to having resorted to poisoning them with chemicals.

That said, the pea aphid is particularly interesting for this story. It has the ability to turn from green in cold weather to bright orange[39] in sunnier conditions through its ability to synthesize the same pigment that makes tomatoes red. This provides it with excellent UV photoprotection. But we have now discovered that the high density of these red pigments also represents a very simple photosynthetic-energy capture mechanism. In a very similar way to how archaea convert light energy, the aphids use the energy to assist their metabolism.

The oriental hornet, *Vespa orientalis*, is another insect full of surprises. It has a bright-yellow band across its abdomen made of tiny grooves and bumps containing layers of pigment that absorb the sunlight. So far, so normal. We know cold-blooded animals can be physically more active when warmed. However, oriental hornets also have a thermal "cooling" system. This allows them to forage in the middle of the day in the desert habitats where they live and where temperatures regularly reach dangerous levels above 100°F. It appears that the bumps on the hornet's exoskeleton provide a thermoelectric current that cools them by 4–11°F. Closer inspection has revealed that this substantial thermoelectric effect provides a mechanism for either heating or cooling the wasp, extending the time of day and seasons in which it can forage. Think of it this way: These hornets use sunlight to pump heat out of their abdomens so that they can forage in hot conditions, when other insects—including their common wasp brethren who embark on most of their sorties in the cool early mornings—cannot. These hornets harness solar-light energy and transform it into another useful form of energy through highly unconventional forms of photosynthesis to provide them with their own air-conditioning.

Figure 5: *Vespa orientalis*—photovoltaic-light-absorbing bands generate electric potential to help cool or heat the hornet so that it can forage for longer than other insects.

The list of creatures that are dabbling with photo-synthesis is not confined to the slugs and bugs we have discussed so far. They also include the North American spotted salamander. Photosynthesizing green algae have been discovered living inside the egg capsules of these amphibian animals in an aston-ishing symbiotic relationship with their embryos.[40] Genetic analysis shows that the algae are fully active, absorbing and using the ammonia waste of the salamander for their nitrogen metabolism. In return, the salamander embryo is detoxified and fed with oxygen.

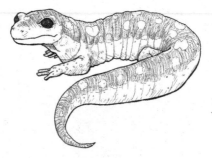

Figure 6: US East Coast spotted salamander *Ambystoma maculatum* carries photosynthetic algae in its eggs. These help to feed and detoxify the developing eggs. The salamander carries the algae in a dormant stage inside its ovaries over a lifetime. This is another example of animals adopting photosynthetic capabilities.

Given that adult salamanders are not transparent, the algal cells stay dormant during their adult lives. When the salamander spawns new eggs, they are fertilized with dormant algal cells from its reproductive organs. Crucially, these eggs are transparent, and the algal cells then start photosynthesizing again. These salamanders only benefit from photosynthesis for a brief period in their life cycle, but the value of this process is such that it makes rearranging their genome to host algae within their eggs worthwhile.

Why these new forms of photosynthesis matter

Ever since the earliest stages of life on this planet, photosynthesis has evolved with it. First, it enabled archaea and bacteria to increase the range and number of environments in which they could thrive. Then, when bacterial cells started incorporating photosynthesis into more complicated host organisms, a complex web of symbiotic relationships formed. Within them, bacterial cells became embedded in a vast variety of hosts, resulting in an enormous expansion of life-forms, survival strategies, and entirely new ecosystems. By working together, they colonized Earth's barren continents, creating entirely new biology-driven geological and nutrient chemical cycles that further increased the exchange of essential nutrients between living organisms and the environment.

Nor was this the end of the story. As we have seen, even animals have begun using photosynthesis to harness energy and enhance their chances of survival. There has been a gradual and subtle transition. Corals are among the oldest animals on the planet. They have had hundreds of millions of years to establish one of the deepest and most successful symbiotic relationships with photosynthetic organisms. Now snails, insects, and even complex vertebrates have also integrated plant genes. Inevitably, we will speculate on where this will go next. Might we see photosynthetic butterflies or bats? Right now, the idea that the process might even include us humans still seems far-fetched, not least because we do not have a lot of surface area relative to our body mass, so it would be extremely difficult for us to absorb the amount of sunlight needed to fuel our intense metabolism. Indeed, the animals described in this section are still the exception to the rule. But, given the amazing ways that photosynthesis has been adapted by different life-forms so far, the never-ending story of evolution means those animals may yet be the beginning of a new chapter.

CHAPTER 4

THE PLANET'S VERY THIN
AND LIVING VENEER

I hope that this book has, by now, demonstrated how photosynthesis is the driver of our global environment. It is the modulator of atmospheric and oceanic carbon. It is the producer of food for us and all other creatures on Earth. It fuels the breakdown of our wastes and the recycling of waste into useful resources.

It seems pretty obvious, then, that photosynthesis is vital for our own survival. If we want to rebalance our planetary equilibrium while feeding a projected population of perhaps ten billion humans, we need to work out how much photosynthesis is happening—and where. We must understand not only how much food we can grow but also how to expand the natural environment that creates soil, generates our oxygen, and recycles the carbon dioxide that is necessary for life to flourish.

That is a tough task. Imagine measuring a transient process going on in the oceans, in Arctic tundra, tropical swamps, and mountain ranges and across agricultural lands in countries that may not be easily accessible or cooperative.

The question of how much photosynthesis is happening is therefore phenomenally tricky to answer. Tricky, but not impossible because, as so often in human science, finding an answer to such a fundamental question about saving life has been made easier by the technology designed by people who spend most of their time thinking up ways of killing us.

Assessments of photosynthetic activity on a global scale have only begun to be possible since the military developed satellites to spy on large and often inaccessible land areas. High-resolution imaging from space started in June 1959 when the US's CORONA or KeyHole strategic reconnaissance satellites were employed to monitor nuclear-missile and strategic-defense installations in remote corners of the Soviet Union and China. It did not take long for defense analysts to understand that they could also monitor agricultural production and whether or not Communist Five-Year Plans were being met. After all, the "fighting readiness" of any military depends on whether the army can be fed.

Soon, the first global climate models were also being

developed to understand the potential fallout and distribution of radioactive dust clouds after nuclear tests. Knowing which way the wind is blowing nuclear fallout is very important indeed. These programs were quickly followed by the LANDSAT program, which could differentiate vegetation from other land and oceanic features, enabling a large-scale global view of forestry, agricultural production, fisheries, and land-use changes.

Eventually, much of the data were declassified for civilian use to make better weather forecasts, manage fisheries, monitor the impact of wildfires, or to trace the global impact of urbanization, infrastructure, and agricultural development. For the first time, scientists could see large-scale natural oceanic blooms in their entirety or measure the difference between green coastal waters and blue ocean water. Red and blue light-absorbing photosynthetic pigments could be traced whether they were in a jungle, coastal ocean, or a cornfield. Best of all, the rise and fall of photosynthesis could be tracked through the seasons as fallow winter fields blossomed in the spring or forests changed leaf color in the autumn.

All these different forms of data enabled us to measure the total amount of photosynthesis happening in the world or its "Net Primary Productivity" (NPP). NPP represents the total amount of photosynthetic activity minus the amount of carbon organisms use for their own metabolism. The NPP is the amount of carbon removed from the environment and stored in biomass in a defined timeframe. We could now discover how much new biomass grows in each separate environment on our planet every year. We can link biogeochemical cycles with the seasonal shifts in photosynthesis. For example, satellites allow us to measure where—and how frequently—the algae that formed the White Cliffs of Dover bloom.

They also fundamentally changed the way we think about the world. Historically, we had thought that the rain forests were the "lungs of the planet," and this technology allowed us to watch how a great green frontier of plants marches rhythmically back and forth across the land masses with the seasons.[41] But, when we were finally able to measure global photosynthesis comprehensively, using satellite data, it challenged conventional thinking. Before, we had seen how the land was green and therefore concluded it was the most productive part of the planet. In contrast, we believed that the much larger blue oceans were relatively unproductive.

Then, in 1998, Paul Falkowski and his team published some findings at Rutgers University that integrated land and marine primary productivity. The result was that we all had to recalibrate a whole range of fundamental, well-established, and comfortable misconceptions.

Falkowski and his team measured the amount of photosynthesis that is happening—the NPP—by how much new biomass grows per unit area per year. Figure 7 shows what they found.

Biogeographic	Aquatic NPP		Terrestrial NPP
Oligotrophic (low nutrient ocean)	11.0	Tropical rain forests	17.8
Mesotrophic (med nutrient)	27.4	Broadleaf deciduous forests	1.5
Eutrophic (high nutrient)	9.1	Broadleaf & needleleaf forests	3.1
Macrophytes (kelp/seaweed)	1.0	Needleleaf evergreen forests	3.1
		Needleleaf deciduous forest	1.4
		Savannas	16.8
		Perennial grasslands	2.4
		Broadleaf shrubs w/ bare soil	1.0
		Tundra	0.8
		Desert	0.5
		Human Cultivation	8.0
Total	**48.5 Gt C/Y**		**56.4 Gt C/Y**

Figure 7: Net gigatons (billions of tons) of carbon biomass fixed by photosynthetic activity per year. Ocean color data (used to quantify marine NPP) are averages from 1978 to 1983. The land vegetation index is averaged from 1982 to 1990.[42]

Look at those figures: The study found that 56.4 gigatons of carbon was being fixed on land but that 48.5 gigatons was being absorbed and fixed as biomass in the oceans. Until then, most people had not been fully aware of the large impact that our oceans had in terms of photosynthetic or biological activity, oxygen generation, and carbon dioxide absorption. It emerged that nearly half of the oxygen we breathe is generated in our oceans and those with intermediate nutrient levels—located between 30° and 60° latitude—were consistently more productive than expected. Kelp and the seaweed gardens hugging continental edges fixed 2 percent of ocean carbon in terms of biomass. In parallel, the open ocean "deserts" like the Sargasso Sea and the Central Pacific Gyre (unfortunately now known as the garbage patch), which had always been considered too far from land to receive meaningful nutrients, were found to be contributing about 20 percent to ocean production.

This was the first time that a truly global perspective of photosynthesis-driven biological activity was assessed. Since then, remote sensing satellites have continuously improved and higher resolution means more precise measurements are being made every year. These have thrown up other positive surprises. For example, re-analysis of recent marine photosynthesis data show us that the surface ocean is more productive than we expected and almost twice as much carbon is exported to the deep.[43]

However, there is an unsettling detail. The historic land-based perspective also seemed to overestimate how much photosynthesis was happening in total. On land, the equatorial rain forests and the northern-hemisphere savannas form two highly productive zones, but large areas of land such as grasslands, tundras, shrubs, and deserts are significantly less productive. Human croplands under active cultivation only accounted for a meager 8 gigatons of carbon fixed out of a total of 105 gigatons of global photosynthetic carbon fixation. If we turned the entire planet into a giant farm, we would dramatically reduce our ability to remove carbon from the atmosphere.

And then there was the question of what happens to that biomass. While the oceans and land turn out to be nearly equivalent in terms of production, the fate of the biomass is very different depending on where it is grown. A cellulose fiber in a tree trunk can be parked there for thousands of years. A microscopic organism floating around a coral reef may be eaten by another creature at any moment. Plants store a large proportion of the energy they have absorbed from the sun in very nutritious carbohydrates, fats, or proteins, which make an excellent meal for other creatures.

We also needed to quantify the biomass that is eaten, shed as dead leaves in the autumn, or buried in the soil. The most practical way to do this was to measure the carbon content as a ratio of an organism's "dry" weight. To do this you must first get rid of all that water. Imagine drying a jellyfish. It is 95 percent water. That means that 20 tons of live, wet jellyfish dries down to one ton of chewy crunchiness. Trees are approximately 50 percent water, which is why we dry our firewood. Humans are somewhere in between, with an average of 65 percent water by weight (in general, children contain a little more water and people with a lot of fat contain proportionately less water. This is because fat contains less water than other tissues).

But tons of carbon still do not represent total biomass. An organism's dry weight is roughly twice as much as their carbon content. Although carbon is a major element in our bodies, it is only one of many. We have calcium in our bones, nitrogen in our protein, and phosphorous in our genetic information. The reason we use carbon as a measurement is so that we can assess how much biomass is passed from one link in the food chain to another. Most of the energy-carrying compounds that compose the food that organisms eat are rich in carbon.

The table opposite summarizes an amazing and heroic piece of work by Yinon Bar-On at the Weitzmann Institute in Israel, which brings together many hundreds of references across multiple scientific disciplines and allows us to see the world in new ways.

Taxon	Example organism	Terrestrial Gt C	Aquatic Gt C	Subsurface Gt C
Plants	Trees	*450*	0	0
Chordates	Fish	0	0.7	0
	Livestock	0.1	0	0
	Humans	0.06	0	0
	Wild Mammals	0.006	0.001	0
	Wild Birds	0.001	0.001	0
Arthropods	Insects/ Crustaceans	0.2	1	0
Mollusks	Snails	0.02	0.2	0
Cnidarians	Jellyfish	0	0.1	0
Annelids	Roundworms	0.2	0.08	0.04
Nematodes	Ringworms	0.008	0.01	0.004
Fungi	Mushrooms	12	0.3	1
Protists	Amoeba/ Zooplankton	1.5	*2*	0
Bacteria	Bacteria	7	1.3	67
Archaea	Archaea	0.5	0.3	7
Viruses	Viruses	0.1	0.1	0
Total		472	6	75

Figure 8: Gigatons (Gt) of carbon by large groups of organisms on land, at sea, and underground. Reported values have been rounded to reflect the associated level of uncertainty. Underlined and italicized numbers are approximations of the biomass of photosynthetic producers (Adapted from Bar-On et al., 2018).[44]

It is fun to spend some time on this table. There is, for instance, more accumulation of carbon (0.2 Gt C) in viruses than there is in humans, livestock animals, wild birds, and wild mammals combined (0.17 Gt C). Marine snails and roundworms also beat the human, mammal, and bird population on planet Earth.

But the most obvious conclusion from this table is that biomass accumulation of carbon happens mostly on land and in trees. No less than 450 gigatons—that's 452 billion tons of carbon—is in our trees out of a total land-based biomass of 472 Gt C. Imagine a food pyramid for what happens on the land. At the narrow top of it are the complex animals with 0.17 Gt C, followed by insects, snails and worms with 0.43 Gt C, then a middle layer of fungi, bacteria, and viruses with 21.1 Gt C. All these organisms are consumers of biomass and, combined, they amount to 22 Gt C. The broad base of the pyramid is composed of plants, whose gigantic accumulation of 450 Gt C feeds all of those consumers. There is a ratio of 20.7 times more producer biomass on land than consumer biomass.

In the ocean, by contrast, there are a lot more animals—including mammals, crustaceans, birds, fish, worms, snails and jellyfish—than there are on land. Add up the carbon accumulation of ocean animals and they total 2.1 Gt C—nearly 3.5 times more animal biomass relative to that on land. As Figure 7 shows, the ocean produces 48.5 Gt C of primary producer biomass compared to 56.4 Gt C on land. The relatively smaller amount of oceanic photosynthetic activity is feeding a much larger volume of animals. It is an upside-down food pyramid where there are more predators than there are food producers. A skinny, little base of photosynthetic organisms is supporting a lot more consumers higher up. This is the opposite of what happens on land. How can this be possible?

Oceans cover 71 percent of Earth's surface and approximately 80 percent of all of Earth's biomass is stored in land plants. In contrast, approximately 70 percent of all animal biomass is oceanic. Or, as Burgess and Gaines from the University of California, Santa Barbara suggest: "Earth has a plant-dominated landscape and an animal-dominated seascape."[45]

The reason for all this can be found in the way different organisms have adapted to their environment. Plants have had to colonize land, which had a relatively low carbon dioxide atmosphere and where the availability of carbon, water, or nutrients was limited. They require roots to draw water and nutrients from the ground, and to anchor them to it. They also need sturdy cellulosic and woody stems to rise above ground and compete for light, while withstanding the wind and the weather. Plants have to grow leaves to absorb as much light as possible, and these must either be robust enough to cope with seasonal changes or replaced on a regular basis. They need to be sufficiently armored to survive persistent attacks from consumers ranging from worms and insects to fungi and mammals. All this must happen over life cycles lasting hundreds, or even thousands, of years. As a result, they are generally large or very large organisms that grow slowly, storing vast amounts of biomass, but passing relatively little of it up the food chain. It means that on land there is a relatively low efficiency of energy transfer—roughly 1 percent—from the bottom of the food pyramid to the next tier.[46]

The oceans are a comparatively carbon-rich environment for aquatic photosynthetic organisms to live in. These organisms have no need for roots, stems, or leaves because they are exposed to currents that transport nutrients over great distances or from the depths. Nor do they need to pump water through their bodies because they are surrounded by it. They have a very high surface-area-to-volume ratio, with small cells readily absorbing nutrients and exchanging waste gases. All this means they have much faster growth rates and need to store virtually no biomass. Look what happens to algae. These are microscopic organisms with life cycles measured in days, and there are many generations of them every year. When they bloom, they can multiply faster than they can be eaten. And, with so many larger predators around, much more of the energy—approximately 10 percent—moves up the food chain.[47]

We can, therefore, conclude that plant biomass

accumulates in large land plants—the primary producers—whereas animal biomass accumulates in large marine animals: the consumers.

Can we use photosynthesis to rebalance our world?

Teasing through these oceanic, land, and ecosystem differences is very important if we are to balance global carbon budgets. Earth has not always operated at the current level of photosynthesis. We know from myriad ice ages that our global atmosphere is directly linked to the amount of photosynthesis and carbon cycling through our biosphere. Too much has in the past resulted in global cooling, while too little means global heating.

For example, during the Carboniferous era, 360–300 million year ago, there was much more photosynthesis than there is now. Enormous organic sediments rained down on ocean and swamp floors to create the vast coal deposits that are the foundation of our industrial wealth. During that time there was so much photosynthetic biomass that the oxygen content of the air was 62 percent higher than it is currently. It meant goose-sized dragonflies—which would never survive in the comparatively thin air we breathe today—could thrive. There was also an average of 800 parts per million carbon dioxide in the air, almost twice as much as today. The atmosphere was perfect for both animal and plant growth.

So much activity meant an enormous amount of biomass was being created, much of it ending up not only in fossilized deposits of coal but also in the oil and gas that we are busy returning to the atmosphere today. Even when oxygen-related photosynthesis inhibition is taken into account, there was a huge win for carbon capture,[48] with three additional gigatons being sequestered each year in the form of vegetation and soil deposition. This is an enormous amount of carbon being removed from the system, and shows what is possible.

Before anyone starts hopping up and down, I am not, under any circumstances, proposing to double the carbon dioxide levels in our atmosphere. Yet it should make us wonder what we can do to make the planet more productive again. Too often in the debate over fossil fuels, there is a sense that energy is a finite resource and that centuries of progress are at risk of being reversed. There are plenty of economists who argue that our current wealth is a temporary blip based on fossil-carbon energy that we will no longer be able to use. They suggest that, in its absence, we will find it increasingly difficult to hold on to all of our economic, social, cultural, and infrastructural accomplishments. Others suggest that instead of relying on energy that has been stored in the ground for hundreds of millions of years, we can save ourselves by using renewable sources. With the exception of tidal and geothermal energy, renewable sources are directly or indirectly based on solar energy. It is, therefore, vital to know if there is enough sunlight energy flowing through our system to compensate for all the fossil fuels we should not use any more, as well as to support economic growth and feed our growing global population.

Of course, there are also scientists who are trying to tap into the sun's riches directly. For example, scientists at Cambridge University are developing an "artificial leaf" that can convert sunlight, carbon dioxide, and water into a carbon-neutral fuel to cut emissions. This carbon-neutral fuel is a liquid, which can be stored, unlike the electrical current from conventional photovoltaic cells.[49]

By the time light hits ocean or land, large portions of the total solar irradiation has already been absorbed by the atmosphere. It is the reason why we position telescopes on the highest and driest mountaintops or move them into space. The radiation that remains after it passes through the atmosphere is still pretty enormous. The solar energy that strikes our Earth is approximately 178,000 terawatts a year. That is 1.78 watts with 17 zeros behind it, or 178,000-million-million watts. In comparison, all of human energy production is approximately 18 terawatts per year. In other words, the seemingly all-powerful human species, with all our fossil, nuclear and renewable resources, creates approximately 0.01 percent of the energy that our planet receives from the sun. And, of course, most of that human-produced energy is derived from plant-based fossil fuels that stored solar energy millions of years ago.

It means there is ten thousand times more energy streaming past us every single day than we currently use. And, if that is not humbling enough, it turns out that photosynthesizing organisms are much, much better at harvesting energy from the sun than humans are at producing it.

Currently, the amount of solar radiation absorbed for photosynthesis is between 3 and 6 percent of the total. If we assume a modest average of 1.7 percent light conversion efficiency, photosynthesizers harvest nearly twenty times more energy each year than humans produce. While it would be impossible to absorb it all, there is still a vast surplus of light energy available to expand photosynthesis, create much more biomass, feed our expanding population, and balance the carbon in our atmosphere again. Fossil evidence from the Carboniferous era proves our world can support a lot more carbon fixation than it does today.

We have the space, the power, and the means to do more than we do. We just need to help our amazing photosynthetic friends, from the small algae to the tallest tree, catch at least some of those extra rays so that—together—we can put it to work for all of us and the planet we share.

After all, we have done it before.

CHAPTER 5

GOOD NEWS . . . AND BAD

The agricultural miracle

Let's have the good news first. Despite all the problems humans have created for this planet, our agriculture has so far, just about—amazingly—kept pace with our exploding population, appetite, and development.

For tens of thousands of years, more people has meant more consumption. After we had eaten almost all the wild animals, we farmed and ate domesticated ones in ever-greater numbers. The soaring numbers of humans over recent centuries has been accompanied by a massive expansion in demand for not only meat but also dairy, eggs, and clothing. All these farmed animals themselves have required feeding—today more than 60 percent of the world's agricultural products goes into nonhuman bellies. Given that 90 percent of the feed energy for warm-blooded farm animals is turned into heat, most of that agricultural output does not get passed on to the ultimate human consumers and is therefore a very inefficient use of these resources. This is causing all kinds of stress and strain to our natural ecosystems.

However, it is worth standing back to marvel at the scale of our achievement in feeding so many large animals. By some measures, we are finding it easier than ever. We used to spend the majority of our time finding, growing, and preparing food. Now it takes less than 10 percent of our time in developed countries and approximately 40 percent in developing countries. Food cost as a proportion of our earnings is, wherever we live, going down as we spend less money on it than in the past. Of course, there are still far too many people even in rich countries who suffer from food poverty or malnutrition. And the news is often filled with deeply distressing images of famine caused by bad weather, drought, war, or disease. But, for most of us, food is available, affordable, and abundant. Feeding billions of people and even more billions of animals is nothing less than an agricultural miracle. Nor is agriculture just about food. We grow wood for structural materials and fuel. We grow herbs, spices, teas, and coffees simply because we like the taste. We grow decorative plants to put in our vases as flowers and medicinal ones to cure us when we are sick. All these activities are based on the systematic improvement of plants for the production of valuable goods. More plants result in more wealth for more people.

None of this would be possible without photosynthesis. Over the centuries, we have refined our relationship with photosynthesizers to make them even more productive for our purposes.

Breeding, irrigation, and bird droppings

These days, very few agricultural plants are still farmed in their original form or, as the advertisements like to say, "as nature intended." You would be hard-pressed to find an orchard filled with crab apples or a field planted with dandelions. Every imaginable plant has been selected, crossed, hybridized, grafted, and bred to enlarge the size of its fruit, increase the yield of its grain, amplify its fragrances, multiply its flowers, or sweeten and soften its flesh. Thinner husks, brighter flowers, more resistance to disease, better storability, or improved hardiness have all resulted from careful selection of the most robust, shiniest, tastiest plants.

Where possible, the growth of cultivated plants has been accelerated, too. The most productive agricultural areas boast multiple crops per year, with mixed legumes followed by hay, wheat, and oats. Another way to increase productivity is, when conditions permit, to grow double crops on the same plot. Beans are planted underneath a canopy of orange trees in the Nile Delta. Sometimes, insecticidal or fruit-bearing shade trees are grown together with tea and cocoa.

Whether it is by traditional breeding techniques, grafting, the careful fertilization of plants with cultivated bees, or the manipulation of sexual reproduction cycles, farmers have continuously improved agricultural productivity. Across entire regions, wild male hops are systematically hunted down and eradicated to ensure that the female plants produce only flowers without seeds. This is because fertilized hop seeds spoil the taste of beer. Hop-hunting is, by the way, a brilliant job. You travel through some of the most beautiful countryside in the world, spending spring and summer working outside—all for the very good cause of beer drinking.

One of the limits to photosynthetic productivity by even the best-bred plants, however, is their requirement for lots of water. Most of the world's drinkable water is used in agriculture, to irrigate enormous areas of marginal land to grow fodder for cattle, sheep, or goats in order to satisfy the human desire for meat. Such water is in scarce supply and our expanding population is making ever-greater demands for it, not only for drinking but also for hygiene purposes and in industry. Still, human ingenuity has been working away for several millennia to find solutions. We have invented all kinds of clever ways to divert, share, transport, and direct water for the purpose of increasing agricultural yields. The oldest cultures learned to irrigate with underground canals, like the Omanis, with their falaj system, or the Romans, with their overland aqueducts. Deep wells, dams, or complex social arrangements to share water are all used to enable the cultivation of unyielding marginal or desert land.

Some of the most beautiful water management systems can be found in the black soils of the Canary Islands, where individual grapevines are cultivated in carefully sculpted dimples of volcanic rock. As the marine morning mist wafts over Lanzarote, it condenses and percolates to the bottom of the dimple, where the thirsty vines make the sweetest grapes. Similarly, a thin strip of land angled just right along the Moroccan coast captures enough of the morning fog to provide water for crops even in the Sahara Desert. Today, where these creative, heroic, or industrial water distribution measures do not suffice, municipal wastewater is being reused to irrigate city parks. And, although desalinating seawater is highly energy intensive, the cost of it is now low enough for it to be used for crop irrigation in Morocco.

Another limitation to photosynthesis is plant nutrition. The first crops were planted maybe 12,000 years ago as hunter-gatherers turned to farming. It did not take long for our ancestors to discover that using animal waste increased the yield of their fields. Animal dung provides nitrogen, organic biomass, and trace nutrients that enrich soil. It improves the capacity of the soil to sustain plants, increases water retention, and feeds myriad soil organisms that in turn improve soil fertility. While poor management of these resources sometimes has unintended consequences, for example polluting drinking water with nitrates, this archaic method is still being used to provide plant nutrition. Today, what is known as multitrophic cultivation of crops—where the waste from one organism can fertilize another—is seen as state-of-the-art agricultural technology.

The enrichment of soil with blood, bones, feathers, and other compostable animal waste is an old tradition that has transformed entire regions. *Terra preta* or enriched "black soil" was produced by South American Indians by the diligent application of manure, fish bones, and ash, as well as vegetable and slaughter remains to make Amazon basin soils into some of the most productive in the world. This is not just a cute ethnographic story. There are estimates that 10 percent of Amazonian agricultural land—an area larger than the state of California—still benefits from this ancient soil improvement. In contrast, the hugely damaging slash-and-burn method, by which rain forest is cut down to create new farmland in the Amazon, produces only sandy soils from which the ash is quickly washed out. These newly depleted soils have very low crop-carrying capacity. It is like burning caviar to toast bread.

Other cultures have long since plowed seaweed into the soil to improve crop production because kelp is rich in potassium. Sometimes, fishmeal, fish extracts, and fish bones were the most practical sources of nitrogen, phosphorous, or potassium to improve agricultural productivity—and these are still being used as agricultural fertilizers now. Rice growers have always known that recycling fish waste increased production and, more than four thousand years ago, the Chinese civil servant You Hou Bin was detailing how to integrate fish cultivation with agriculture.

The best-known marine fertilizer was first documented in 1609, when Garcilaso de la Vega described how the Incas "use no other manure but the dung of sea birds, of which large and small varieties occur on the coast of Peru in such enormous flocks that they seem incredible to anyone who has not seen them." Guano, or hardened and mineralized droppings of seabirds, became a global industry in the nineteenth century when the explorer Alexander von Humboldt, in South America to measure ocean currents, became fascinated by the activities he observed on Peruvian guano-loading piers. Realizing that what they were transporting was regarded as something much more valuable than bird excrement, he sent samples back to Europe for chemical analysis. The subsequent papers Humboldt wrote describing the virtues of guano were widely read. He effectively launched a global fertilizer boom in which guano suddenly became one of the world's most valued commodities. When it ran into short supply, people began desperately to look for an alternative. When natural nitrate deposits—so-called saltpeter—were discovered in South America, the trade in them became so lucrative that it even triggered a war between Chile and Peru.

This was the time of the Industrial Revolution and, without a dramatic improvement in agricultural productivity, moving the expanding human workforce from the fields to the factories would have been much more difficult. And, throughout the nineteenth century, the need to extract ever-greater amounts of food from plants was turning agricultural research into a new science. Agronomists like Jean-Baptiste Boussingault carefully—and remarkably accurately—quantified the demand for nitrogen fertilizers. They showed that high-protein crops, such as pulses and beans, which are rich in nitrogen, require more nitrogen-based fertilizer than comparatively low-protein barley, for example.

On his country estate in Victorian England, John Bennet Lawes and his partner Joseph Henry Gilbert experimented with chemicals to boost cabbage and turnip production. Lawes discovered that he could extract fertilizers from bones with sulfuric acid to produce a more soluble "super-phosphate" fertilizer. He patented his invention in 1842 and then built the world's first commercial fertilizer-producing factory. His subsequent wealth and estate flowed into the UK's agricultural research station, now called Rothamsted Research. It is the longest-running institution of its kind anywhere in the world. And yet, in spite of all such advances, agriculture was still heavily reliant on limited amounts of guano and saltpeter. And, by the end of a century that had seen such expansion in the wealth and number of human beings, scientists had good reason to be very worried about what the next one had in store.

Bread and bombs

On a stiflingly hot summer night in 1898, Sir William Crookes rose to deliver his inaugural speech as the new president of the British Academy of Sciences to a packed Bristol music hall. He was the inventor of the Crookes Tube, which paved the way for the cathode ray tubes used later in televisions. He was also the discoverer of thallium and an enthusiastic spiritualist. His perspiring audience that night was given a grim prophecy for the future, "a life and death question for generations to come." He warned that "as mouths multiply" and "food sources dwindle," people would begin dying of hunger in large numbers around the 1930s. "England and all civilized nations are in deadly peril of not having enough to eat."

He predicted that when the mountains of bird guano and Chilean saltpeter had all been mined, crop yields would plummet, and people would starve. Sir William, who was far from immune to the imperialism and racism of other Englishmen at the time, went on to describe "how the great Caucasian race will cease to be foremost in the world, and will be squeezed out of existence by races to which wheaten bread is not the staff of life."

Clearly, the prospect of food shortages also applied to people dependent on rice, corn, and millet but, if Sir William's diagnosis was tainted, his proposed remedy was prescient. "It is through the laboratory," he said, "that starvation may ultimately be turned into plenty . . . It is the chemist who must come to the rescue." The answer, he suggested, was to find a way to convert some of the nitrogen that represents 78 percent of our atmosphere into artificial chemical fertilizer so that we can "make bread from air."

Among those inspired by this challenge was a Jewish-German scientist called Fritz Haber. Today, his name is inextricably linked with war crimes, mass murder, and the kind of evil-genius scientist caricatured in films. There is no simple explanation for his behavior. He had troubles growing up: His mother died giving birth to him and he had a fraught relationship with his father. Yet, he was apparently a happy, bubbly child raised by his loving stepmother and three devoted stepsisters.

At the age of thirty-two he pursued and married the truly formidable Clara Immerwahr (whose last name literally means "Always True" in German). She was the first woman to earn a PhD in Germany and, armed with her magna cum laude degree, had just commenced some ambitious experimental research when she was whisked off her feet—literally—by Haber at a dance class. Initially, this fiercely independent women's rights campaigner rejected his marriage proposal so that she could continue her studies. But she eventually succumbed to his charm, married, and produced a son. Although Clara could no longer balance her activism and academic career with motherhood, the Habers maintained a grueling social calendar that featured a glittering array of academic superstars, including Lise Meitner, Otto Hahn, Hans Geiger, Gustav Hertz, James Frank, and Albert Einstein.

A gifted physical and electrochemist, Haber strove to meet Crookes's challenge by developing an industrial method for making "bread from air" based on atmospheric nitrogen fixation. But his efforts were soon redirected. Unfortunately, nitrates are not just used to fertilize crops; saltpeter was also commonly employed in the production of explosives. Therefore, in times of war, nitrogen-rich minerals that could otherwise be used to fertilize food production were diverted to make explosives. Indeed, bat droppings found in caves provided much of the firepower for the American Civil War and the foundation of the DuPont industrial fortune.

At the start of the First World War in 1914, Germany needed an unprecedented amount of exploding nitrogen compounds because the British navy's blockade had halted the import of saltpeter and nitrogen from South America. So Fritz Haber immediately volunteered to develop "war chemicals" for the military, and soon proved himself to be both creative and productive. He proposed ways to use smaller amounts of chemicals and find synthetic methods to provide a substitute for Chilean saltpeter. In his workaholic way, he then expanded his war efforts by conceiving, developing, and weaponizing gases such as chlorine and carbonyl chloride—also known as phosgene. He personally supervised the burial and triggering of deadly asphyxiating gas bottles in Ypres, famously shouting, "God punish England!"

When the newly decorated captain of gas warfare returned home to celebrate his battlefield success, his wife confronted him in the presence of their illustrious guests. She was outraged and insisted that chemical weapons were a "perversion of the sciences, corrupting the discipline that should confer new insights into life." That night she took his service revolver and shot herself in their garden. Her thirteen-year-old son found her as she was dying.

It is impossible to imagine what motivated Fritz Haber the next morning. He abandoned his traumatized boy and his dying wife. It was more important for him to return to the front so that he could direct more gas attacks. By the end of the war, one hundred thousand soldiers had been killed and a million wounded by his chemical warfare. This terrible history was compounded by the work Haber did between the First World War and his death in 1934. He developed cyanide gas to eradicate pests in grain silos. Along with his colleagues, he promoted the development of a product called Zyklon. Later, this became the basis for Zyklon B, the mass-extermination agent used in concentration camps to murder many millions of people, including members of Haber's own extended family.

None of us can, or should, overlook what Haber's wife had called "perversions of science." But this is not all of his story. For a decade before the war, Haber had worked on fixing atmospheric nitrogen for making agricultural fertilizer by synthesizing ammonia. He understood the enormous energy requirement, low yield, and astronomical costs entailed in chemically forcing nitrogen and hydrogen atoms together. He tried to solve the reaction kinetics involved in this unlikely chemical reaction. His team sorted through three thousand catalysts that dramatically increased the yield of ammonia production until they reached commercially feasible levels. Not satisfied with having solved the fundamental research challenges, he then worked with Carl Bosch on the engineering challenges of building efficient chemical reactors for industrial production. This involved creating new chromium- and nickel-enriched steel alloys that could withstand the corrosive nature of highly reactive hot hydrogen under pressure.

These accomplishments led to Haber being awarded the Nobel Prize in Chemistry in 1918. It was a controversial decision, given that this was the same year peace had come at the end of a war made so much worse by his poison gas. The Nobel Prize committee chose to overlook his toxic war record because of the scale and consequence of his achievements. To this day, the Haber–Bosch process is the main method of producing ammonia for fertilizers. For more than a century, mankind has continued to increase the application of his invention year by year, with ammonia production rising an average of 3.5 percent annually to more than 150 million tons.

About half of our world's ever-increasing population of humans are fed because of Haber's synthetic fertilizers. For all the pain and devastation he wreaked elsewhere in his life, Fritz Haber was a scientist—perhaps *the* scientist—who prevented mass starvation with his discovery of how to "make bread from air."

Feed the world

The question of whether we can produce enough food to feed all of us, however, is one that will continue to be posed for as long as there are enough of us left to ask it. Little more than sixty years after Sir William Crookes's speech in 1898, the global population of humans had doubled in size to around three billion. That was despite two world wars, the Spanish flu pandemic, and a series of terrible famines. Indeed, the prospect of the latter was a constant threat for developing countries where birth rates were exploding, crop disease was rife, and farmers were struggling to adapt traditional methods. In some cases, Fritz Haber's synthetic fertilizers were causing further problems because tall wheat stalks collapsed under the weight of their nitrate-inflated heads at the first puff of wind.

This was when another scientist by the name of Norman Borlaug stepped forward. He was a farmer's son who came from humble origins in the American Midwest and had initially failed his entrance exam for the University of Minnesota. Eventually, he got accepted to its College of Agriculture's forestry program, but even then he had to pay his way by working for the Civilian Conservation Corps. He worked at the CCC alongside starving men during the Great Depression, an experience he later said had profoundly affected him. "I saw how food changed them," he said, "all of this left scars on me."

After the Second World War, he was appointed to help the Mexican government improve wheat production. Over the next sixteen years, he crossed more than six thousand wheat isolates as he developed smaller, hardier, higher-yield species that did not get knocked down by wind or bad weather. By carefully selecting breeding locations across Mexico 700 miles apart, he took advantage of two separate seasons to test many more wheat crosses every year. He outmaneuvered plant pathogens so that he always had a strain available that was resistant to disease. By 1963, 95 percent of the wheat planted in Mexico was based on Borlaug's strains and the country's production increased sixfold.

At the same time, India was seeing its crop production plummet. The situation worsened after war with Pakistan in 1965 and the Bihar Drought a year later. Fearing famine and millions of deaths, the Indian government sought advice from Mexico on how to improve food security. Borlaug went to work, overcoming extraordinary bureaucratic, logistical, banking, and cultural hurdles to send hundreds of tons of seeds specially selected to suit the climate of the Indian subcontinent. At one point, his wheat seeds were held up in port by the racially charged Watts Riots in Los Angeles. When they eventually arrived, despite being over-fumigated and germinating poorly, Borlaug's seeds were so successful that schools had to be shut to store record harvests.

Borlaug became known as the "Father of the Green Revolution"—a revolution that from 1960 onward increased wheat yields 2.5-fold in the world's least-developed nations. He found out that he had been awarded the 1970 Nobel Peace Prize while working in his experimental Toluca wheat fields outside Mexico City. It is estimated that he prevented the starvation of more than a billion people over the course of his life's work.

At the beginning of the industrial revolution, the idea of a global population of 7.7 billion would have seemed not only ridiculous but terrifying. In 1804, the population of the world was only a billion, and there was relatively little livestock. Dropping an additional 6.7 billion people on the planet—not to mention billions of farmed animals—would have utterly wiped out all ecosystems. Hordes of desperate people would have mowed through every resource like a locust plague, sucking the marrow from the bones of the last sparrow. Billions of humans would have starved.

But, starting with the global use of Peruvian bird droppings 215 years ago, we have managed to double our population over and over and over and over again—four times in total.

Years to double population	Year	Human Population
	1600	500,000,000
204	1804	1,000,000,000
123	1927	2,000,000,000
47	1974	4,003,448,151
51	2025	8,110,000,000
111	2100	10,880,000,000

Figure 9: UN World Population Prospects: the 2019 revision, median projections.[50]

In the last forty years, we have also managed to lift billions of people out of poverty, educate more children than ever, reduce famine, and sustain a global middle class who live in the kind of luxury that would never have been dreamed of by even the most pampered emperor a few hundred years ago.

Sure, the planet is groaning and creaking under the burden of vast environmental damage being done by our species. But the good news is that we may not need to perform this colossal population-doubling miracle again. The total number of humans in the world is expected to peak, according to the median United Nations forecast, at 10.9 billion in 2100.

There are many indicators suggesting we might not even reach that level. Hans Rosling, the late Swedish statistician, has shown how the impact of better child survival rates, the provision of basic health care, and female education can result in a much lower overall birth rate.[51]

Over the last four hundred years, humans have repeatedly reshaped the world to overcome ignorance, hunger, and poverty. We can yet amplify those achievements through education and health care to address the challenge of feeding the world, without killing the planet in the process.

As we have seen, the use of sunlight by plants exceeds many times the sum of all human energy production. Furthermore, it is evident that much more plant life is possible on our planet, and that there is more than enough sunlight available to increase the production of photosynthetic biomass to feed our peak human population. This increase in productivity has to be accomplished in a way that does not obliterate remaining biological diversity and resources. That is not just the right thing to do, it is also almost certainly the only way we will thrive as a species, because on this shared planet we have a shared set of interests.

If we want to feed ourselves, we will need to feed the rest of the living world, too.

Why wild animals matter

Now for some bad news. Even the most committed skeptic would probably acknowledge that humans are abusing and polluting our planet. We see it every day in our own lives as we drive in our cars, tip plastic into our bins, and order packaged food. We are bombarded with news of environmental catastrophe. We watch nature programs about rain forests, oceans, and exotic animals that are being destroyed, degraded, or made extinct. It is not difficult to believe that we might share the fate of the animals that are no longer with us. Mammoths, saber-toothed tigers, geoducks, American lions, dire wolves, and giant sloths did not survive the last ice age, possibly because humans did. We hunted, ate, and then dressed in the skins of these animals to stay warm. Many of the large mammals that have survived include rabbits, wolves, and cats that since then have been domesticated by us and made into pets or livestock.

The chart opposite summarizes the impact humans have had on global animal wildlife. Based on fossil records and global system extrapolation, Yinon Bar-On and his team, who created Figure 8 in the previous chapter, estimate that the ascent of man has resulted in an 82.5 percent descent in the total of wild animal biomass, a lethal trend of destruction that continues today. We are not, however, destroying all animal life. As the chart shows, the world's livestock animals carry 2.5 times more carbon today than the total population of wild animals did 100,000 years ago. On top of that, there is also 0.06 gigatons of human carbon—this is 50 percent more than the total for wild animals when we were just starting out after the last ice age. The combined total of human and animal biomass today is 4.2 times bigger than it was before human life. In other words, we have extracted enormous value from the natural world to transform it into livestock and human animal carbon.

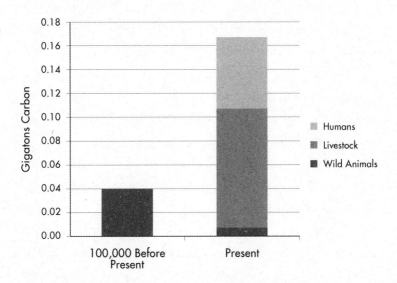

Figure 10: Global animal biomass 100,000 years ago compared to today. This shows that humans have increased total animal biomass through livestock breeding, at the expense of wild animals. Redrawn from additional reference material (Bar-On et al., 2018).[52]

After we hunted and killed most of the great land mammals, we turned to those in the oceans. In the nineteenth century, tens of thousands of boats hunted whales for the lamp oil that lit the homes and factories of a young United States. Herman Melville's 1851 novel, *Moby Dick*, tells the story of Captain Ahab's maniacal pursuit of one whale in particular. It is shocking to read, not least because so little was understood about the biology, ecology, or lifestyle of a species—described by Melville as a "spouting fish"— that was being hunted to the point of extinction.

When I was a boy, advancements in whaling technology were still seen as a sign of industrial progress. Whales were tracked with sonar and shot with explosive harpoons fired from powerful cannons before being hauled into factory ships, where they were sliced up with huge chainsaws. Eventually, films of desperately fleeing whales with their lungs exploding in obscene fountains of blood prompted an outcry and the very slow enactment of international regulation to protect them.

While people have clamored to "save the whale" over recent decades, fish stocks have become progressively stressed, over-hunted, or destroyed. Even some of the most environmentally aware of us are perfectly happy to consume healthy, easy-to-digest, rich-in-omega-3, low-cost fish. We have created aquaculture to supplement wild fish; indeed, today, the total amount of farmed fish exceeds wild-caught fish. At the moment a large proportion of farmed fish is actually fed and fattened up with fishmeal made of wild-caught fish. We continue to hunt, devour, and exterminate this last vast resource of wild animals just as our ancestors did with those tasty mammoths.

Perhaps because fishing fleets are throwbacks to our hunter-gatherer origins, they are often seen as virile symbols of nationhood. They get heavily subsidized or ferociously defended by the political leaders of different countries. Even today, enormous energy is invested in defending territorial rights to protect ever-fewer jobs to catch ever-fewer fish with some of the most sophisticated technology ever invented. It means that in all practical ways, we have already accepted the extinction of blue-fin tuna. We watch television shows celebrating those who have caught the highest-prized fish where their value is driven by their scarcity. Headlines are made when a single tuna fish fetches $3.1 million at an auction, and investors are planning for the day when prices rise even higher by purchasing the best specimens to store in ultra-low-temperature freezers. In a world with low interest rates and low-yield bond markets, banking on the disappearance of this much sought-after species is a lucrative business. Catching and eating the few remaining specimens of these magnificent top predators of the oceans is a good story for the media and good business for investors. It is akin to celebrating the shooting of the last cheetah, speculating in rhinoceros meat futures, or eating dodos. Oh wait, we have already eaten all the dodos.

On one level, it might be possible to argue that the extinction of a particular species does not matter so much in the great scheme of things and, in any case, has little to do with photosynthesis. But if we care about the survival of our own species, we should be deeply concerned about the survival of others. That is because the largest driver of environmental destruction and the extinction of other animals is habitat loss. When wild animals become rare or extinct, it is a measure of how much we are extracting resources from our planet's most productive ecosystems for agriculture, cities, and industrial production. And, above all, when we destroy these natural habitats of coral reefs, savannas, and forests to create shipping lanes, fields, and roads, we also destroy the systems that absorb carbon.

The opposite of an ice age

Since the beginning of life on Earth, the capacity to harvest sunlight and produce biomass has, apart from during the ice ages, generally increased. Until we came along.

Humans are having as much impact on the carbon cycle as an ice age, except that we are taking the planet in the opposite direction. Even as we pump more heating gases into the atmosphere, we are reducing the capacity of the planet to absorb and balance these emissions.

We are cutting down tropical rain forests at an insane rate. We are draining and drying swamplands to make new farms. We are burning woodlands to make fertilizer or space for ranches and crops. We are obliterating coral reefs for the construction of industrial harbors or shrimp farms. We are destroying our oyster banks and estuaries with poorly controlled run-off from industry or agriculture.

This catalog of environmental destruction seems as inevitable and irreversible as it is familiar.

The table below shows the Net Primary Productivity (NPP)—the net amount of carbon removed from the environment and stored in biomass—over the period of a year by various habitats. The American ecologist Peter Stiling compiled this data in 1996, in the last years of land-based assessments of NPP (before we were able to use satellites). It is important because, although satellites are fantastically accurate at providing detailed biomass measurements of a particular mangrove swamp or estuary for example, they don't yet provide us with an understanding of how these ecosystems compare with each other, and specifically with cultivated land.

Ecosystem Type	Area (10⁶ km²)	Mean NPP (g/m²/y)
Tropical rain forest	17	2,200
Tropical seasonal forest	7.5	1,600
Temperate evergreen forest	5	1,300
Temperate deciduous forest	7	1,200
Boreal forest	12	800
Woodlands and shrublands	8.5	700
Savanna	15	900
Temperate grasslands	9	600
Tundra and alpine	8	140
Desert and semidesert	18	90
Extreme desert and ice	24	3
Cultivated land	14	650
Swamp and wetland	2	2,000
Lakes and streams	2	250
Open ocean	332	125
Continental shelf	26.6	360
Algal bed and reef	0.6	2,500
Estuaries	1.4	1,500

Figure 11: Annual Net Primary Productivity (NPP), in grams of carbon fixed per square meter per year (g/m²/y) until 1996, by ecosystem type.[53]

The problem is that the highly productive natural habitats listed in the table above are being replaced by ones that are less productive. With the inexorable increase of industrial and urban developments, we are forcing overall NPP to reverse.

What were vital carbon-absorbing lands have been turned into net carbon emitters. Many of the most productive habitats are being replaced by agriculture. On average, current farming practices are only fixing 900 grams of carbon per square meter per year, yet the rain forests they replace would have removed 2,200 grams of carbon.

In other words, to feed the world we are converting our most productive environments into ones that are less productive. We are impoverishing the photosynthetic capacity of the planet, which is the start of all food chains, in order to feed ourselves. We waste natural resources, prevent natural habitats from regenerating themselves, and consequently lose more species—animal and plant—every day. We are throwing away thousands upon thousands of years of ecological enrichment in return for the sugar rush of this year's food crop.

Our industry, housing, agriculture, and pollution are now extending into the highest mountain meadow, the coldest frontier, and the deepest forest. As they do so, the fingers of human progress are unpicking the natural safety mechanisms that have stabilized the interconnected environment since the last ice age.

Look closer at the table. It includes eight million square kilometers of semi-frozen non-agricultural tundra. For tens of thousands of years, hardy arctic and alpine plants absorbed carbon and built up biomass during summers. Small plants grew, slowly accumulated carbon, and then froze where they died. This trend began to reverse in the 1970s. As the lakes and soils thawed, the decomposing biomass released carbon dioxide and methane—flipping vast tracts of land from being carbon-absorbing into carbon-emitting. Methane emissions matter, because their increase means that over the next decades, they will contribute an additional 50 percent of atmospheric heating relative to carbon dioxide.

The US Department of Energy's Oak Ridge National Laboratory has conducted a multi-year study of northern peatland ecosystems called Spruce and Peatland Responses Under Changing Environments (SPRUCE). Peatlands represent approximately 3 percent of land mass. In carefully controlled enclosures, Paul Hanson and his team showed that even a small degree of warming results in overall drying of the peatland, turning it from a carbon sink into a net carbon emitter.[54]

Previous chapters of this book have explained how the volcanic ash that dusted glaciers and the evolution of snow algae helped our planet to escape "Slushball Earth." Today, dark particles of air pollution settling on snow and ice have the same sunlight-absorbing effect, but this time we are not in an ice age and really do not need warming up. The reduction of light reflected back into space—the albedo—is speeding the thaw of Earth's cold places and keeping scientists awake at night. In early 2020, the melting of New Zealand glaciers accelerated when they turned a caramel color from smoke blown across the Tasman Sea from Australian forest fires. As the Arctic ice sheets disappear, the ocean that is exposed beneath them is darker and absorbs more sunlight which, in turn, accelerates the thaw. A warmer Arctic means a warmer tundra, which makes more wildfires with more ice-blackening soot, which melts the ice all over again.

The collateral impacts of warming, fires, and poor land management are all reducing the capacity of environmental systems to absorb atmospheric carbon dioxide and are accelerating us toward "tipping points." Once these are reached, we will no longer be able to reverse the impact of the destructive consequences of our carelessness. Reducing emissions will no longer suffice to decrease warming and environmental degradation.

We have set in motion a series of vicious cycles that spiral downward into deeper and deeper environmental degradation. For instance, poverty drives farmers to send their goats into ever-more remote areas where they will munch their way through the bushes and shrubs that bind a thin layer of soil together. When this vegetation has gone, the soil will wash away in the next rainstorm, leading to more poverty and more desperate farmers. In Brazil, allowing the destruction of the Amazon is a tried-and-tested strategy to win elections. As I write this, the rain forest is burning at an accelerated rate to clear land for cattle ranches. Soon, we will have burned enough forest to stop the trees from creating the rain that is needed to feed the cattle. The destruction follows a familiar pattern whether the motive is to plant crops, herd cattle, or dig for gold. First a road is built. This road then sprouts a network of smaller paths that in turn lead to clearing larger areas. The trees come down, and when the land stops producing, we cut some more trees.

And before we try to shovel responsibility for this catastrophe on to the sagging shoulders of unscrupulous cattle ranchers in the Amazon, we in the supposedly environmentally conscious developed world should accept our own share of it. Look at us as we push trolleys around supermarkets, selecting our aesthetically perfect shrink-wrapped produce while the crooked-shaped vegetables and smaller apples are deemed by customer preference to be fit only for animals. In the 1960s, American consumers spent 18 percent of their income on food; today it is barely half of that.[55] This trend can be seen worldwide. We want to pay less money to eat more, while still wasting and throwing away much of what we buy. The same agricultural miracle that was described earlier in this chapter has led many to become obese. Since 1960, our world population and food production has more than doubled. At the same time, the number of obese people has exceeded the number who are undernourished.

The same process that is filling human bodies with cheap calories is subjugating ecosystems to industrial agriculture, with terrible, planet-changing consequences. The use of Fritz Haber's invention of fixing nitrogen into fertilizer with ammonia is itself among the largest drivers of carbon dioxide emissions. The breakdown of this fertilizer on agricultural fields is also one of the largest producers of nitrous oxide—laughing gas—which absorbs more solar energy than carbon dioxide and has 298 times more atmosphere-warming potential than carbon dioxide over the next hundred years.

Synthetic fertilizers, improved agricultural production techniques, and high-yield breeds were not the only factors that enabled the agricultural miracle of the past sixty years. Another key part of the chemistry that made this possible was the development of pesticides.

Seeds, fields, feed-lots, processing plants, and food waste attract a lot of hungry creatures. Insects, fungi, bacteria, viruses, mollusks, birds, and rodents all want to share in the sudden abundance of ready-to-eat food. Catastrophes such as the Great Famine in Ireland were initiated by a fungus called *Phytophthora infestans,* better known as "potato blight," which was probably imported to the island by ships bringing guano from Peru. Our modern, mechanized, large-scale, global agriculture, which sustains 7.7 billion people, would be impossible without the use of chemical agents—insecticides, herbicides, fungicides, nematicides, and molluscicides—to kill insects, plants, fungi, worms, and snails. Even if we chopped down every last forest and plowed up every nature preserve, there would not be enough land to produce the food we consume today without pesticides and chemical fertilizers. The average loss in food production due to pests is estimated to have been around 20 to 40 percent per year since the 1960s. Without the use of pesticides, the losses would have been between 34 and 37 percent higher,[56] the equivalent of being unable to feed 2.7 billion people.

The problem is that these chemical pesticides obliterate a lot of nature in their path, particularly if they are misused. The use of one herbicide—atrazine—is so widespread that there are traces of it in the snowflakes of Antarctica. In global breadbaskets like Brazil, where there is reckless application of pesticides, entire communities have been poisoned and schools sprayed with chemicals from planes. The litany of chemical damage and abuse could fill libraries as farmers, under relentless pressure to produce more with less, further industrialize production across the world.

Just as it would be wrong to outlaw the use of pesticides altogether, it would equally be a mistake to wipe out insects that are needed to pollinate crops and other plants. Bees, for instance, are in such steep decline that commercial farmers have resorted to loading hives on to trucks and transporting them long distances to pollinate monocultures—often in a fog of chemicals—before offering them a meal of substitute sugar instead of honey. Insects, in general, are in rapid global decline, which is having a residual effect on other animals. For example, numbers of sparrows, which feed their young with them, have dramatically decreased.

An enormous proportion of the most fertile agricultural lands are no longer increasing their crop yield. There are myriad reasons for this, including soil loss, salt accumulation, urban encroachment, and lack of water. Too often we are killing or poisoning the very parts of nature that would help regenerate soils or pollinate our crops. We are destroying not only our environment, but also our capacity to produce food.

Until a few thousand years ago, the Rub' al Khali, or the Arabian peninsula's "Empty Quarter," was filled with lakes, lush vegetation, and water buffalo. Now this most arid and lifeless of deserts serves as a warning to us all about what may happen to even our most productive land. It should show us that the fate of our oceans, rain forests, and glaciers is not some distant event from the past or a mere possibility about the future. These environments are being wrecked right now, even when we are able to monitor the impact with satellite images and can study data models showing the consequences. The destruction is happening as you read this sentence.

Why politics matters, too

What can we do? Prioritizing natural ecosystems over human life to let people starve would be grotesque, inhumane, and unjust. The work of Norman Borlaug demonstrated that starvation can be prevented and measures to intensify agriculture could, other things being equal, feed a larger population.

But everything else is not equal. Many famines are driven by terrible policy choices. How, for instance, can millions of people be starving in an oil-rich and verdant country like Venezuela? There are countless well-entrenched and widely practiced misalignments of what we know and what we do. Today, governments across the world are pandering to protectionist interests that exploit the environment while promoting unsustainable agricultural practices with subsidies and counter-productive regulations.

In the UK, for example, unrelated to avian flu or other disease management, widescale slaughter of wild birds is considered to be appropriate to protect agriculture and absurd hunting practices. At the same time, millions of European dairy cows are subsidized to the tune of $1 per day, while a billion people in the developing world live on less than that. The 1.5 billion cows used in global agriculture each emit between 150 and 260 pounds of methane[57]—which has an atmosphere-warming potential eighty-four times that of carbon dioxide over the next twenty years. If cows were a nation, they would be third—behind only China and the USA—in a league table of the world's methane polluters. The problem is so severe that many research teams are now examining ways to alter cows' gut organisms to reduce or stop their methane emissions. These include feeding them a type of seaweed that reduces some of the gut microbes or even seeding clean calves, born by caesarean section, with alternate bacteria to replace the methane-producing gut organisms.

Saudi Arabia irrigates enormous areas of its desert with roughly 10 cubic miles of ice-age water every year to grow potatoes and alfalfa for air-conditioned cattle. If we are not turning agricultural crops into fodder to feed animals, we make it into low-cost fuel to feed cars. We are building cities in the desert and then demanding energy-intensive air-conditioning. We drive ever-larger vehicles, travel more frequently and ever farther. We are desperately bailing out airlines so they can go back to pumping out the same amount of carbon into our atmosphere as they did before the COVID-19 pandemic.

And, if all this was not bad enough, we also subsidize the fossil fuel industries with trillions of US dollars a year[58] and then let our politicians complain that they do not have resources to develop sustainable alternatives. Every year we invest immense amounts of money to promote and encourage the use of non-renewable resources that we know are harmful. It is no use blaming bad industrial practices in the past. We are doing this to our planet and ourselves right now. Oil production is still increasing and will continue to increase for at least another ten years. Global post-tax subsidies for fossil fuels, meaning the cash incentive plus the environmental damage they cause, are $5.2 trillion a year.[59] That is $5.2 million times a million. Imagine a world where we had $5.2 trillion to invest in sustainable technology—every year.

But even if we managed to decarbonize all human activity, we would not fix our current environmental challenges. In addition to reducing our carbon footprint we also must increase the capacity of our planet to reabsorb the carbon we have released.

That means using the existing agricultural land more effectively while protecting the unfarmed wild places and natural systems surrounding us so that they can sequester carbon, create oxygen, and support plant and animal life.

In the Nile Valley, for example, irrigation has increased primary production to 1,200 to 1,500 grams of carbon per square meter per year, one of the highest NPP rates anywhere in the world. This improvement is due to a combination of modern agriculture techniques, with clever double cropping and multiple harvests per year on the rich Nile Delta soil, compared to seasonal agriculture based on annual floods. Sadly, it is still the exception[60] but it shows that it can be—and has been—done.

Instead of promoting and financing destructive and unsustainable farming practices such as ripping up mangroves, cutting down rain forests, or exhausting our soil, we should make sure that the global annual budget of $700 billion for public environmental subsidies is targeted more effectively to increase support for non-agricultural land development. One recent report from the 2019 Food and Land Use Coalition suggested that retargeting only a fraction of this would "generate a societal return of around $5.7 trillion annually."[61] Land use changes alone would produce socio-economic benefits, including better human health, accelerating income growth for rural populations, the creation of 120 million rural jobs and improved biodiversity, while restoring 3 billion acres to nature. Now imagine the impact of redirecting vast destructive practices, like fossil fuel subsidies, into creating more and more sustainable wealth.

Humans have shown they are clever enough to

transform the world, reduce suffering, improve living standards, and live longer, while still not running out of food. However, in doing so, we have also been releasing vast amounts of carbon into our atmosphere and destroying the fundamental capacity of our planet to regenerate itself. It goes without saying that we need to reduce carbon dioxide emissions rapidly and dramatically. But this is not sufficient in itself. We also need to increase the ability of our ecosystems to fix the mess.

Fortunately, we know how to do this: Photosynthesis, which made the planet, can save it again.

CHAPTER 6

HOW PHOTOSYNTHESIS CAN SAVE THE WORLD

The bottom line—or "what's in it for us"

Humans are both the epic heroes and the arch villains of our planet's story. If this were a movie, the obvious development in the storyline would be for our species to reach a fork in the road, one where our true destiny is finally decided. Will the dark side prevail as we choose short-term growth, material comforts, and financial gain at the expense of our planet's health and the lives of future generations? Or will we save the world by turning to the light and Earth's source of life?

The preceding chapters have outlined biological solutions to grow more food, increase the global photosynthetic productivity that captures carbon, as well as expand the total living space and volume of biological activity on our planet.

All of us are also aware that such a narrative is usually played to a soundtrack of impatient sighs from policy makers and economists. Who, they ask, will pay for it all? They tell us that planting a trillion trees would require a different kind of tree—one of the "magic money" variety. They point to polls and election results that show that few people want higher taxes, or explain how producers have little incentive to dispose of all their waste responsibly. Saving the planet, they worry, will simply "cost the Earth."

In the face of such cynicism, we will not get very far by pointing out that all the resources we use and every life we live is a priceless gift from the planet. Instead, the way to win this argument is to demonstrate how by creating renewable resources and providing more food, new jobs and wealth can be created. After all, people are better stewards of things they value and tend to neglect those they regard as fanciful or uneconomical.

What if we can show the hardest-faced capitalist that we do not face a binary choice between our ecology and the economy? What if we can prove that—in addition to being the source of all life in our world, the food we eat, and the oxygen we breathe— photosynthesis is the key to reviving our economies and maybe even making investors richer?

The first step in this is to start putting a price on the damage we will do if we carry on as we have been.

Back in 2006, Lord Nicolas Stern wrote a seven-hundred-page review for the UK Government that tried to assess the economic impact of carbon pollution.[62] The report concluded that if we were to carry on unchecked, it would make the world economy 5 percent poorer, and there was a reasonable probability this cost would rise to 20 percent. These are enormous numbers that have the potential to grind global human productivity into relentless decline. A 5 percent reduction would mean losing $4.4 trillion, an economic shock comparable to the COVID-19 pandemic happening every year. A 20 percent reduction would be equivalent to wiping out the entire US economy. Significantly, the Stern report approached this challenge as a market failure that could be remedied with assertive, rapid action to change behavior. He argued that the cost of reducing emissions could be limited to around 1 percent of the world economy through new taxes and charges, based on putting a price on carbon-intensive activity.

Predictably, the Stern report prompted a stream of criticism from interest groups and industry lobbyists, as well as hedging from economists and hesitation from politicians who rarely need an excuse for more prevarication. In 2016, ten years after his original report, Stern returned to the fray to declare that if anything he had been too conservative in his assessment of the damage being done. "With hindsight," he said, "I now realize that I underestimated the risks. I should have been much stronger in what I said in the report about the costs of inaction."[63]

Since his report, no less than fifty-seven different varieties of carbon pricing initiatives have gotten underway[64] at regional, national, and supranational levels. None of them is perfect and local circumstances will continue to change. Depending on what time frame each operates in, the prices per ton of emitted carbon dioxide that are being discussed range from $15 to $80. It is good that different approaches are being tested. A range of methods for fairly distributing the resources gained from taxation across income, class, generations, and those who stand to lose from decarbonizing society are being evaluated. Imagine how the economics of constructing a building would change if every ton of cement cost an extra $80, while simultaneously every ton of wood would create an $80 credit for storing away carbon? Cement would soon be decarbonized and a new boom in wooden buildings would materialize almost overnight.

There is another reason why such carbon economics are important. To give an example, the most pressing concern for a farmer in Madagascar is how to feed his children. If he is to stop burning rain forest, he needs an incentive to become a better steward of the environment.

The value of storing carbon or enhancing nature has to have a price that can be linked to income for those who will be most affected. And for this to happen we will need to change the way our economies operate.

One of the problems, however, with current measurements of value or cost is that they can confound and confuse as much as they communicate. Although figures for tons of carbon emitted are hugely important and increasingly understood, they represent daunting and highly impersonal numbers that are difficult to think about or feel. And, as we have seen, although a reduction in carbon emissions is necessary, it is not going to be enough on its own to rebalance our planet.

In short, estimating the cost of damaging behavior is not the same as working out what all our planetary resources—the air, fresh water, food production, medical therapeutic herbs, and entertainment—are worth. It is the question they always ask in business: "What's the bottom line?" In this case, it is notoriously difficult to answer. Where do you draw the boundaries? Is fresh water worth $73.48 trillion, as suggested by the UN Environment Programme World Conservation Monitoring Centre?[65] In which case, how much value is there in Antarctica's frozen plateau or Greenland's ice sheet? What price do we put on network services, like a bee pollinating a small shrub that serves to feed a goat, when that same bee gets eaten by a lizard and the shrub also provides shelter for a turtle? Indeed, what is each breath we take worth—and does the value vary at different times of our lives? The London Zoological Society and the World Wildlife Fund estimate that nature underpins $125 trillion in economic activity annually. Australia's Great Barrier Reef alone contributes $5.7 billion, while supporting 69,000 jobs.[66] Meanwhile, the BBC Earth Index assessment suggested that a single beaver provides $120,000 worth of annual wetland management and forest cultivation services, while dung beetles, which improve the nutrient quality of soil by burrowing and burying dung, are collectively worth some $380 million per year.[67]

Whatever the exact figures, few people would contest that nature is worth a whole lot more than anything that we, either as individuals or collectively, could ever earn. If we lose land, habitats, and species, we destroy the very things on which wealth creation depends. It is akin to driving a golden egg-laying goose to extinction. That is, in anyone's book, extremely bad economics.

We should be able to agree that metrics matter, even if we cannot put an exact price on the contribution made by every living thing. Measurements tell investors, policy makers, and companies what needs to change, as well as showing us where they need to be held to account. Indeed, there is an old proverb that states: "Be careful what you measure, because you will get more of it." Measuring the living world may be the best way to ensure its survival.

The trouble is that there is an alphabet soup of acronyms, representing different organizations trying to measure nature, which make the whole subject rather hard to digest. For example, the WMO (World Meteorological Organization) operates the GCOS (Global Climate Observing System) and came up with 50 ECVs (Essential Climate Variables). On the other hand, we cite NVDI (Normalized Difference Vegetation Index) numbers to describe how much vegetation (or algae in my ponds) we can see with our remote sensing satellites. And if that's not enough, how about a dimensionless quantity that characterizes how much leaf coverage exists in plant canopies known as the LAI (Leaf Area Index)?

Fortunately, there is one unifying measure that takes all of these and more into account. This is the Net Primary Productivity, or NPP, that we have already discussed in previous chapters. It corresponds directly with the value of photosynthesis, without which there would be virtually no life or ecosystem— and definitely no economy, no money, and no people.

NPP is the net production of organic biomass from inorganic sources per unit area for a given period of time. It is a measure that allows us to compare a coral reef with a marginal desert, or the value of using land for cattle grazing instead of growing trees. If we incorporate carbon into soil, bury it in construction material, or store it in spent oil wells, we can test these changes by the same measure. We can contrast the carbon impact of recycling a plastic bag with the carbon capture of a birch tree's growth. No matter where we are in the world or what we are studying, we can use a single unified measure of how many grams of carbon have been absorbed a year by biomass for each square meter.

NPP has another important advantage: It is relatively easy to measure. Indeed, measuring it can be as simple as holding a ruler at arm's length at different times to assess how much a tree has grown. You do not need an MBA or a PhD. And it's not only easy to do, it is also emotionally accessible. Because all of us can visualize the size of a square meter and estimate how much biomass there is in it; we can work out just how much stuff we need to grow.

Above all, measuring photosynthesis through NPP gives us a common reference point to manage carbon taxes, plan for land use changes, and balance both sides of the carbon-emission-versus-absorption equation. For instance, it may show that a logging company—vilified by environmentalists despite replanting more trees than it cuts down—is significantly more sustainable than a green-friendly "vertical city farm," which uses vast amounts of energy while producing relatively little. The "net" in Net Primary Productivity means that both the carbon inputs and emissions are captured by any assessment on the impact of a decision. It allows us to consider the value of steel relative to wood, concrete, or glass. Measuring it by square meter means we can compare the carbon management performance of small countries with gigantic ones.

The only problem is the term itself. "Net Primary Productivity" is about as emotive as a bag of sand. The measure of photosynthesis should be a call to action that reconnects us to the land and sea of our fragile world. At great risk of ridicule, my personal suggestion is that we should rebrand it a "Vitality Unit" which could be shortened to the acronym ViU. Even the hardest lumberjack could assess the way he logs an industrial forest to maximize vitality; and the most dry-as-dust accountant could find a way of pricing vitality units or assessing the kind of stimulus needed to promote biomass growth and to make investments greener, lusher, and more productive.

The rest of this chapter therefore gives some examples of how to soften our environmental impact and restore vitality to the planet.

They do not require a supernatural force, technological somersaults, or impossible amounts of money. Nor do they mean we have to walk around barefoot wearing hairshirts. Most of the following examples have been demonstrated already. Many of them are creating new jobs, businesses, and fortunes right now. All of them will increase the vitality of our planet and our global level of NPP through much, much more photosynthesis.

Bigger crops and "pastural care"

In the past few years, an army of small agri-tech start-ups, university labs, and industrial producers has been mobilized to grow more food on land that is already farmed. Taking advantage of incentives and tax breaks from governments, a new wave of entrepreneurism is washing over agriculture. Experts in remote-sensing satellites are helping improve "precision" farming, leaving no area in a field undercultivated. Robotics scientists are finding ways to be economical with the application of pesticide and fertilizer while still improving yields. Data are being mined and artificial intelligence designed to recognize the happiest, most productive plants.

Agri-tech is very hip right now because financial accelerators, incubators, investment funds, engineering companies, and pension funds have spotted that fertile ground is, well, fertile ground for making new money. Historically, oil and gas exploration attracted the smartest geologists, nuclear power groomed brilliant physicists, while financial service quant shops vacuumed up math whizzes. But in recent years, agricultural universities that were once seen as academic backwaters have been attracting some of the sharpest minds to help develop "digital agriculture" with microwave- and laser-pulsing "swarm farms" employing the "Internet of Agricultural Things." Wild acronyms for crop and soil health protection, animal welfare, agricultural, and precision engineering are twisting excited tongues as new economic opportunities and industrial strategies are developed.

One of the priorities must be to solve our dilemma about insects. For decades we have been using chemistry systematically to exterminate anything that might eat our crops. This approach, which usually means the use of millions of tons of pesticides, not only threatens the future of a whole host of other creatures—birds, bats, and amphibians—that feed on them, but also kills the insects that play such a crucial role in pollinating plants. In other words, we risk killing the little creatures that are needed to help grow our food in the future and also feed wildlife. Fortunately, science is now exploring new and better ways to protect agriculture. Experiments with natural pheromones to distract moths or ants from crops have been shown to be highly effective. Biologists are manipulating the immune systems of pests to make them less harmful. We now have field trials that interfere with beetle genes so they cannot digest—and no longer want to eat—potato plants. And more attention is being paid to traditional methods, for instance planting stinky marigolds next to bean plants to keep root-eating worms away.

Another area where there is both an imperative and an enormous opportunity to do more over the next few decades will be in preventing the loss of our most fertile soil. The Midwestern states of the US have been one of the most productive agricultural areas on the planet but, since the 1850s, nearly 9 inches out of a total of 14 inches of their fertile topsoil has been lost.[68] That soil was a precious gift passed down to us from the ice ages that deposited ground loam where once there had been mountains. Almost two-thirds of it has already washed away and, if soil loss continues unchecked at the present rate, all of it will have been squandered by 2107. According to a United Nations–backed study, "a third of the planet's land is severely degraded and fertile soil is being lost at the rate of 24 billion tons a year."[69] It is already one of the chief reasons why wheat production in France, Germany, and the UK has stopped increasing despite improved farming practices such as precision agriculture that would have otherwise increased yields.[70]

The good news is that better farming practices can prevent such soil degradation. The ancient art of terracing hillsides can help. So too can planting in curves rather than straight lines, using a technique known as contour plowing. Simple solutions, such as barrier bags in drainage ditches or shallower plowing of fields, have both increased yields and improved soil in many countries. Combining or alternating crops with potatoes and legumes have similarly reduced run-off and soil loss.

We know what we need to do. Development aid programs are helping share experiences and the best practices. Big producers, including China and Brazil, have begun, belatedly, to manage their agricultural soil loss (although unfortunately, as we have seen, Brazil is simultaneously destroying its rain forest, leaving a legacy of unproductive sandy soil). Other countries are faring better; for example, Ethiopia alone has rehabilitated 17 million acres—or 27,000 square miles—of its precious land by protecting the soil from erosion with hardy plants.

The Loess Plateau in China is the size of France. It was one of the cradles of early civilization and as recently as five hundred years ago was still a fertile area. But overgrazing by sheep and goats, deforestation, and crop-planting on slopes stripped it of most vegetation. The result was that enormous quantities of its light topsoil washed into the Yellow River, turning it—in fact—yellow, and by 1994 the area was declared a desert. An extraordinary repair project has seen a series of terraces built downstream to stop the silt washing away. This is then used to grow corn or vegetables. The steep valley slopes have been stabilized by the planting of fruit trees, dune grasses, and bushes to keep the wind from blowing away the soil. What was desert is now filled with greenhouses, fields, and forested mountainsides supporting twenty million people, whose incomes have tripled and health improved.

Sometimes, the oldest tricks are the best—the promotion of traditional agricultural practices is proving to be good both for communities and the environment. As mentioned in the previous chapter, *Terra preta*, or "black soil," was formed between 2,500 and 1,000 years ago in the Amazon basin and is still yielding significantly higher crop yields than the surrounding land. Ancient Amazonian societies made it by plowing charcoal ashes, fish bones, pottery shards, vegetable matter, and organic waste from their kitchen dumps into the ground around their villages. This resulted in soils that remain fantastically fertile even after hundreds of years of farming. Encouragingly, some of the big phosphate and fertilizer companies are learning from this and using crop waste (known as biochar) and other organic matter to upgrade African soils. This will not only produce more food but also enable us to store more carbon in soil.

We need to go even further, and soon. For centuries, agriculture has become ever-more mechanized, industrialized, and automated with the aim of increasing crop yields to feed many more people. Soil loss and salinity, resistant diseases and pests, as well as alternating droughts and floods, are all neutralizing the benefits of modern agriculture. We have reached the point where further mechanization, more inorganic fertilizers, and pesticide-resistant crops will not work.

Instead of industrializing agriculture, we need to nurture better land stewardship. This does not have to be confrontational or challenging. Supermarket chains already insist that their suppliers grow consistent and safe produce, for example by requiring farmers to have larger borders and hedges surrounding their fields to separate forests and farms, thereby producing better food while also protecting the forests. Large commodity companies are rewilding unproductive land in the US Midwest to turn them into buffalo grazing grounds again. Supermarket chains and commodity producers can work with farmers across the world to use their land in better, more sustainable, ways. Government incentives and subsidies should be refocused around rewarding those who create new jobs in rural areas by increasing soil carbon, soil fertility, and soil respiration.

Soil can be upgraded and used even in the heart of our cities. Mexico City has a scheme to turn all of its food waste into highly valuable soil carbon that dramatically reduces the emissions from municipal rubbish and will help feed the city in the future. Similar projects elsewhere are extracting phosphates from wastewater sludge instead of flushing them into rivers. Urban waste can be a valuable resource to reclaim land, create arable gardens to feed the poor, and clean up some of our cities' worst slums. In Nairobi's infamous Kibera, Dhaka's Korail, or Addis Ababa's Yedekeku Menderoch, "Extended Metabolism Models" are showing that even the lowest-income communities can organize to improve their environment. Chengdu in China, home to 14.5 million people, is turning itself into a vast sustainable garden to manage run-off and heat. Singapore is promoting rooftop farming not least because the extra soil and vegetation reduces floods and cools the city in summertime. Creating diversified and often healthy foods in such urban gardens may not end world hunger, but it certainly helps. One study puts the potential output of city-grown vegetable crops at 180 million tons.[71]

Another area where technology, government intervention, and market economics is accelerating improvement is in reducing the amount of energy used to grow our food. Although crop cultivation remains one of the largest sources of carbon emissions, overall energy efficiency is getting better in trading areas such as the EU.[72] One of the reasons for this is that energy consumption in agriculture is no longer subsidized. Consequently, farmers who might never think of themselves as part of a green movement are looking to save money by using more renewable energy or turning animal waste into bioenergy. A report by the European Union in 2015 outlined the full life cycle energy demand of growing, processing, shipping, packaging, consuming, and disposing of key foodstuffs such as butter, oranges, and olive oil.[73] This showed energy consumption per cultivated acre has decreased steadily since 1990.

In contrast, even the current version of the Build Back Better bill—which, at time of writing, is struggling its way through the US Congress, with its $555 billion of incentives focused on stimulating a greener economy, reducing consumption, and increasing sustainable energy production—does little to revise agricultural incentives for corn-based fossil fuel production or to reduce the fossil fuel reliance of modern agriculture. The carbon footprint of US agriculture is an untapped opportunity.

Ammonia synthesis, which has given us artificial fertilizer and enabled much of modern civilization, also creates massive carbon emissions and is the single largest energy consumer in the chemical industry. For each of the 175 million tons of ammonia produced in 2016, two tons of carbon dioxide were emitted.[74] The new holy grail in chemistry, therefore, is to find a way of smashing nitrogen into hydrogen to produce ammonia without immense pressure and heat. The search has already seen exotic and creative experiments being conducted with microscopic spikes of carbon that physically catalyze the generation of ammonia,[75] copying the way bacteria use enzymes to extract nitrogen,[76] as well as deploying a series of chemical reactions at lower temperatures and pressures.[77] Whoever succeeds in finding a scalable and efficient way of producing ammonia at lower cost, in terms of both money and carbon emissions, will not only make a fortune but also go a long way to making food production sustainable over the next century.

One of the biggest changes we need to make in the way we farm is to find a less destructive form of grazing. This is a book extolling the benefits and beauty of photosynthesis, so it should not exactly come as a shock to discover that I am critical of our livestock farming methods. After all, the way goats, cattle, and sheep munch their way through vegetation means they are hardly promoting plant growth. Worse still, they also reduce soil cover and accelerate desertification.

Before human carnivores start howling in protest, I should explain I am not opposed to meat production. In fact, pastureland with its grasses, shrubs, flowers, mosses, and clover can—if managed properly—contribute to the biological resource cycle in a constructive way. There are species of strong, hardy grasses that have evolved to withstand grazing teeth with horizontal stems called rhizomes that spread outward underground and are very good in terms of NPP. Prairies, savannas, and grasslands are often described as "upside-down rain forests," because of the biodiversity and complexity of their underground root systems. These hardy grasses can rapidly grow back, flower, and seed themselves once animal herds have migrated to other feeding grounds, not least because animal excrement, which has already helpfully broken down the nutrients in fibrous cells, is so readily available for plants to reabsorb. Indeed, these grasses do better when they are grazed because, if they are left alone, their woody stems grow so tall they fall over, covering the soil, locking up nutrients, and obstructing further plant growth by blocking the sun. "Ungrazed" grasslands and pastures have lower NPP than those where animals are allowed in to eat them. These ungrazed grasslands often become net carbon emitters if not managed effectively.

Like all things in modern agriculture, however, grazing has to be balanced and managed properly. The best new methods include tight herds that are shepherded in a confined area for twenty-four hours, where they can "take half and leave half" of the plant life before being moved to a neighboring plot. Cameras and computer chips can stop herds getting mixed up, so this method can work just as well in wildlife reserves or on common lands as it does in fields with fences. Herding animals may become a modern profession again.

Controlled grazing not only dramatically reduces soil erosion, but also improves the diversity of species living on the land. As in all biological systems, predators increase plant productivity. This is true of both the smallest spiders providing the soil with nitrogen through the insects they eat and the largest bears fertilizing forests with the waste from the fish they have caught in rivers. The reintroduction of wolves in some areas has helped the turnover of nutrients between plants and animals by keeping the numbers of their natural prey in check.

Together, science and nature can repair our land. On the one hand, we have precision agriculture engineers measuring levels of soil respiration to enliven them again. On the other, there is this great diverse range of animals, spiders, soil crustaceans, and diverse plants which, if allowed to live in our farmed fields, can all protect the crops and help to feed us.

Across the world, there are 10 million square miles of grasslands that could increase NPP and grow more food if they were actively managed. In some places, pastures are being abandoned because occupations such as sheep-farming are no longer viable. Annually, an area of land the size of the state of Oregon is being handed back to nature because people have begun to prefer synthetic fabrics to wool. This is a huge opportunity, and much more land could either be freed up for more productive arable use or properly rewilded. Farmers, foresters, herders, and shepherds could be paid to be stewards of the environment, or to measure and monetize an increase in NPP. Crucially, all this is quite labor-intensive, so it could create a lot of new jobs and new incomes that would also make some of the poorest areas of our planet better off. In the US, there are multiple large-scale rewilding programs sponsored and supported by organizations like American Prairie, the Rewilding Institute, Defenders of Wildlife, the Sierra Club, and the Great Plains Restoration Council. Even national nuclear research facilities such as Fermilab at the Argonne National Laboratory have operated a Prairie Restoration Project, where local seeds were used to remove agricultural crops and weeds to restore the original prairie grass–dominated landscape. Through ranch purchases, collaboration with state and federal agencies, and careful selection of marginal wilderness areas, large tracts of land can be connected to enable wildlife ecosystems to reconnect and re-establish a foothold across states and boundaries. For example, since 2004, American Prairie, together with federal and state agencies, has protected and reconnected more than 5,000 square miles of land—an area roughly the size of Connecticut—to restore biodiverse wilderness.

A revolutionary new product

Rather than burning down rain forests to make room for more cattle, a smarter investment for the health of both bank balances and the planet might be to put money into the hottest new agricultural product: wood.

Around 385 million years since the first tree grew on this planet, the material they produce has become fashionable again. Wood is not only renewable, it is also practical, flexible, and affordable. Materials-science experts are beginning to think about how wood can be used to make cars, clothes, and even biodegradable glue. Crushing bamboo with great pressure turns it into fireproof, hard, machinable flooring. Prefabricated homes and commercial buildings, high-rise towers, and elegant Nordic architectural fantasies are being assembled using new forms of processed wood that combine quality, strength, and stability. Forest Green Rovers, an English League Two soccer team, is even building its new stadium—Eco Park—entirely out of wood.[78]

This is the beginning of a new way of working with wood. We must have been through a similar process at the dawn of the Iron Age. Who would have thought the brownish-grey dirt of iron ore would be much use for anything? But centuries of ingenuity and experimentation have seen it processed to make steel for thick bridge-spanning girders or the thinnest of surgical scalpels. In the same way, wood is now being broken up into its components and processed to make everything from transparent materials to replacements for plastics. Although some of these products are not yet cost-effective, they have already spawned an entire research industry looking into ways of making materials that are lighter, more flexible, and more reusable than plastic, metal, or glass. Densified "super-wood" is as strong as steel and six times lighter. Wooden buildings save 10 to

30 percent on heating and cooling energy. And all of this is to say nothing about how brilliant wood is for storing away carbon for centuries. The building sector alone could bury 10–68 million tons of carbon per year in modern timber buildings.[79]

In anticipation of a new wood-powered industry, there has been an increase in long-term investments in trees, and resource managers are looking for new places to grow forests. Companies that own forests and manufacture timber-related products are being bought by pension funds, charitable trusts, and universities attracted by the idea of sustained growth and predictable returns.

In both financial and environmental terms, trees appear to be a better bet than other alternative crops. Some investors are looking to make money from so-called BECCS—Bioenergy Carbon Capture and Storage—schemes. These require organic biomass to be grown, dried, and burned for energy production without the ashes and carbon dioxide being emitted into the atmosphere. Where there is excess biomass, such as from forestry or building material waste, BECCS can be useful. But it is very important that these sorts of schemes do not come at the expense of productive land by competing with agriculture or destroying ecosystems. Often biofuels, including those made from corn and ethanol, are worse for the environment because they use up land needed to feed people, while the carbon waste is neither sequestered nor stored.

Plant more trees, everywhere we can

Most trees will never be as telegenic as giant redwoods or as productive as those in tropical rain forests. But they can all play a part. The tilting, "drunken" forests of the tundra or the scrawny shrubs on savannas still add up to billions of tons of biomass, as can be seen in the table below.

Region	Type of forest	Million trees	% of total	Ground-based observations
Tropics	Tropical coniferous	22.2		0
	Tropical dry	156.4		115
	Tropical grasslands	318		999
	Tropical moist	799.4	43%	5,321
Northern	Boreal forests	749.3		8,688
	Tundra	94.9	28%	2,268
Temperate	Mediterranean forests	53.4		16,727
	Montane grasslands	60.3		138
	Temperate broadleaf	362.6		278,395
	Temperate conifer	150.6		85,144
	Temperate grasslands	148.3	25%	17,051
Other	Flooded grasslands	64.6		271
	Mangroves	8.2		21
	Deserts	53	4%	14,637
Total trees in billions		**3,041.2**		

Figure 12: Number of trees in different living spaces based on satellite observation and thousands of ground-based validations.[80]

Before we industrialized the planet, there were six trillion trees. Today we have half that number left—around three trillion. Trees account for the vast bulk of living biomass, storing no less than 279.6 billion tons of carbon. They also support many other ecosystems, make rain, and protect the land. Sadly, on average, we destroy ten billion trees each year without replacing them. In 2019, forest fires in Siberia, Australia, and the Amazon burned more than 26 billion trees.

A lot of this damage can be undone. In Africa, the saplings that spring back from the root networks of felled trees are being allowed to grow as part of a natural regeneration project for forests in Ethiopia, Burkina Faso, Niger, and Senegal. This has led to 240 million new trees in West Africa alone. By balancing the shade and shelter of forests with the production of nuts, fruit, and firewood, these reforestation programs have largely been self-sustaining. Of course, these multinational projects are complicated and difficult to implement in remote and poor countries. Yet, remote sensing and other assessment tools can also serve to accelerate support for the network of local projects.

In addition to letting trees grow back in formerly forested areas, there are savannas, semidesert, cities, road boundaries, and public lands that provide an enormous amount of potential planting space. In just one day—July 11, 2016—a vast army of 800,000 volunteers across India planted a then-record of 49.3

million trees along roads and railways. This was part of the country's commitment to the Paris Climate Accord, which will see it spend $6 billion reforesting 235 million acres—or about 29 percent of its territory—by 2030. In 2019, Ethiopia smashed India's record by planting 350 million trees in a single day to leave a "green legacy" and reverse a process that has seen forest cover in the country fall from 35 percent in the 1890s to just 4 percent today. Approximately half of that loss was between 1990 and 2010. The plan is to expand this program to four billion trees by encouraging every single citizen to plant at least forty seedlings, with local initiatives selecting species that will thrive in their areas.

Bhutan, a tiny mountainous country whose population is threatened by floods from thawing glaciers flushing down steep canyon walls, has prescribed a minimum forest cover in its constitution and has a target to offset 50 million tons of carbon emissions. Its much larger near-neighbor, Pakistan, has unleashed the "Ten Billion Tree Tsunami Programme" that will involve planting an additional 850,000 acres with trees and enabling natural regeneration of previously damaged lands. Many countries are picking up shovels to plant seedlings today that will create environmental benefits for decades to come.

Such projects are creating new jobs and commercial opportunities, while improving the recreational value of the land. Trees can grow for centuries, all the time transferring 69 percent of the carbon they capture into the soil. In the US, the state of New York has set itself a goal of ending deforestation by 2030 and planting at least 865 million acres of marginal, degraded, or unproductive land with trees. Even Donald Trump signed an executive order to commit the US to the One Trillion Trees Initiative launched at the World Economic Forum. Given the losses being sustained, we should aim higher. If every person on the planet planted 153 trees over the next 25 years, we would get to four trillion before 2050. Even my 83-year-old mother with her bad knee manages to promote the growth of about 1,100 new trees every year.

There is more than enough space for them. For instance, there are 27 million square kilometers of temperate grasslands, semidesert, and desert that could be more productive. The African Union-led Great Green Wall initiative is creating a 50-kilometer-deep wall of forests and rangelands to keep the Sahara Desert from expanding. Twenty seven million square kilometers is an area three times bigger than the US. Just increasing the average productivity of these lands by five grams per square meter would result in 130 million tons of additional carbon being sequestered, the equivalent of all Brazil's annual carbon dioxide emissions. To give you an idea of how viable this is, tropical rain forests have an NPP of

2,200 grams per square meter per year. The potential to increase the ability of our planet to absorb more carbon is simply enormous.

Jean-François Bastin, an ecologist and a geographer, has used Google Earth[81] to show that there are 2.2 million acres of land available for reforestation "outside of urban and agricultural areas"—in other words, 3.5 million square miles of land that would naturally support woodlands and forests if it had not been logged or degraded. It is land that is primed for tree growth and, if forested, would mean 200 gigatons of carbon being bound in increased biomass— enough to absorb two-thirds of the excess we have in our atmosphere today. Significantly, Bastin's study was based on 78,774 separate sampling points, each with ecologically relevant recommendations on what grows naturally in that area, making it a tool by which local people can take decisions on reforestation independently of governments and corporations.

Imagine what we could do if we got the other 7 million square miles of the desert and semidesert to grow more biomass, or what the potential of improving the coastal ocean areas—a further 10 million square miles—could be for global NPP and vitality. The point is that there are vast areas on this planet including oceans, cities, and cultivated and uncultivated land that can, driven by photosynthesis, produce more life than we do today.

Greening coastal deserts and other seaside tales

This is where the story about photosynthesis gets personal for me. Before I start, let me sound a note of caution. Greening the Saudi Arabian desert or Libya's Al Khufrah oasis sounds wonderful until you realize that the supply of fresh water, as much as 25,800 million cubic meters of Ice Age water per year, or 26 cubic kilometers of ancient water. It is estimated that by now, natural Saudi Arabian aquifers have been depleted by 80 percent.

Producing food in the desert with *sea*water, however, is a very different matter. For the past ten years, I've tried to prove that we can do just that by turning my academic obsession with algae and photosynthesis into a business. My company seeks to convert coastal deserts into some of the most productive land possible. We pump seawater into open ponds, where we grow local microalgae to make biomass and high-value products such as pigments or vegan egg-white substitutes. Almost all the inputs are free. Instead of pumping expensive carbon dioxide or nutrients into our ponds, we use the seawater-based phosphate, nitrate, and carbon dioxide that are available naturally. We grow local algae year-round as they would grow naturally during their seasonal coastal bloom. Our energy is generated entirely from the wind and sun, and as a result, our carbon and freshwater footprints are unbeatable.

It has been an exhausting journey. Our current site is 335 miles from the nearest city where critical supplies can be sourced. We have been battered by sandstorms; our seawater pipeline has washed back up on shore, twice. In addition to COVID-19, everything else that can go wrong has gone wrong. And yet, the company has grown from working with small Plexiglas ponds in South Africa to growing areas 260 times larger in Oman and 15 times larger again in Morocco. Our recipe has been simple: Use what nature provides. The larger the ponds become, the more stable operations are. By working with local algae, which are perfectly adapted and acclimatized to the testing desert conditions, we are certain to produce—whether it's a cold coastal South African winter below 43°F or a blistering Omani summer higher than 120°F. Our other secret has been to work with local people who know how to get the best out of their local coastal desert environment.

It is an "additive capacity" that creates primary productivity on land—and not to mention work for the local population—where it did not exist before. Importantly, this "new" primary productivity does not remove resources from other organisms and does not deplete the environment of living spaces for other species. Our ponds absorb the necessary carbon out of the coastal water, thereby deacidifying it, so that it can then be used to grow new plant biomass. Our approach is about as far as one can get from the familiar safety and stability of labs, and that has one enormous advantage: There is no limit to the scale at which we can grow algae, as long as we have access to empty land and can reach clean seawater with our pipeline.

We estimate that of the 6 to 7 million square miles of hot desert worldwide, covering an area larger than the US and Europe combined, approximately 200,000 square miles have the appropriate properties that could be developed for this form of coastal seawater production. Scale matters: A global network of pond operators has the potential to draw down 2 billion tons of carbon dioxide per year. That is 20 percent of the excess carbon dioxide we are emitting globally, annually. This biomass can be buried at low cost, to remove carbon dioxide and to rebalance our planets with the underutilized powers of the ocean. Of course, we cannot do this alone. However, together as a global network of pond operators, we can significantly contribute to decarbonizing the atmosphere again.

Even more ambitious is the plan by Anastasia Makarieva and Victor Gorshkov, two Russian physicists from the Petersburg Nuclear Physics Institute, for new coastal forests to create their own rain. Their analysis[82] shows how coastal forests could trigger cloud formation that would in turn stimulate more forest growth and start a virtuous circle of bringing marine atmospheric moisture on to land. These new forests would not only provide freshwater to the desert and absorb carbon, but also offer coastal regions protection against devastating floods, droughts, hurricanes, and tornadoes.

Since the publication of their original paper in 2010, new atmospheric models have been tested. It is now thought that large coastal forests can draw ocean moisture deeper inland, protecting against desertification, creating productive agricultural land, reducing temperatures, and making large areas more habitable. Certainly, there are challenges, such as finding enough fresh water to start the process. These are solvable problems, though. The largest desalination plant in the world is currently being built in Morocco to produce water for agriculture in exactly such a coastal area. It is powered with locally generated renewable energy.

There are also multiple initiatives underway to make conventional crops more salt-tolerant so that they grow on low-grade soils. Prickly pear cactus is a well-known desert crop that is salt-tolerant, requires little irrigation, and can live in very poor soils. Blue agave, which is the source of tequila, is a very efficient plant that again can create valuable biomass on desert land while also feeding bats and pollinators and enriching soil. Trees such as pin oaks, honey locust, or eastern red cedars can be grown on relatively salty, marginal lands. It is interesting to see how illegal Afghani opium poppy farmers have found ways to work entirely off grid, using solar power to irrigate large areas of deserts and turn them into highly productive land.

In the meantime, many coastal areas, including some of the world's largest cities and agricultural deltas, are threatened by rising sea levels. The Netherlands, which has more experience in building effective flood defenses than any other country, is preparing to abandon large areas of land in the face of a sea level rise of two meters or more this century. And yet natural systems such as mangroves, coral reefs, or oyster beds can stabilize and protect coastlines from storms, as well as restore ecosystems to their original productivity. In many places, they can grow fast enough to provide this protection even with rising sea levels. For example, in spite of many scientists' concerns that corals would submerge faster than they could regrow, so far at least, undisturbed reefs have exceeded the vertical growth rate necessary to compensate for the current sea level rise. Not only do oyster beds provide coastal protection but also an opportunity to trap sediments needed to build new land.

The really interesting development here revolves around the unloved mangrove swamp. It is estimated that mangroves reduce the costs of floods by $82 billion each year in China, the USA, India, and Mexico alone. Mangrove forests form in a tropical belt of 25° latitude either side of the equator. They provide very strong armor to protect coastlines from erosion because their complex root systems trap sand drifting in coastal currents and then slowly grow outward, increasing the forest size as they create more land.

In spite of the poor level of nutrients in the marine or brackish coastal flats where they grow, mangrove forests also have unusually high photosynthetic productivity, because they recycle the few nutrients extremely efficiently through the ecosystem they foster. They store 69 million tons of carbon per year, equivalent to annual emissions from 25 million US homes. What's more, because of the particular way they grow, mangroves trap a further 296 million tons of soil carbon—equivalent to the emissions of 117 million US homes. These amazing trees manage to thrive in seawater by producing a waxy cork that coats their submerged roots and excludes up to 97 percent of the salt from their surroundings. It also means their roots are often starved of oxygen. So, in response, mangroves have evolved a way of storing gases when they are submerged during high tide, providing the soil's bacteria with air and protecting its natural nitrogen compounds. The whole system provides enormous living spaces for tropical species, sheltering at least 170 species of large marine animals as well 23 trillion young fish and 40 trillion crabs, shrimps, and mollusks. Above the waterline, mangrove canopies support swarms of insect-eating and fruit-dispersing birds.

Despite these contributions to the diversity and vitality of our planet, mangrove forests are still routinely being stripped away to make room for shrimp farms or palm oil plantations. The result is highly impoverished production systems that expose local populations to storms and destroy natural coastal fisheries. Preserving and expanding the 50,000 square miles of coastal mangroves has to be one of the most effective ways of increasing our planet's productivity. It is also cheap and easy to do because effective planting strategies have already been developed. Often, all that is needed to enable mangroves to grow is to put up a simple fence around them to protect them from grazers.

A few miles out from the shoreline, further avoidable environmental devastation is happening every day. Humans continue to build ever-larger and more technologically sophisticated boats to catch dwindling stocks of ever-smaller fish. The "bycatch" (fish that were not the intended target) are thrown back into the sea, where they usually die. The nutrients they once consumed are increasingly being mopped up by blooms of jellyfish. Once again, we know what we need to do. And in very many instances this means that we simply back off and let nature repair itself.

Increasingly, fish—together with fishing jobs—can be protected, managed, and preserved in designated marine sanctuaries. These are areas with corals, kelp beds, and other marine hotspots where excessive fishing, boating, diving, and destructive recreational activities are prohibited. Often, they are large sections of coastlines or island chains, but some are pockets surrounded by commercial fisheries, while others are so inaccessible that they have had little direct human contact. From Scotland to Antarctica, these sanctuaries have been stunningly successful, allowing many endangered or presumed-to-be-lost species to rapidly recover, corals to grow again, and new economic value to be created for sustainable development.

Between 2006 and 2016, President George W. Bush and his successor, Barack Obama, collaborated to create the 580,000-square-mile Papahānaumokuākea Marine National Monument between Hawaii and the Midway Atoll. This sanctuary protects more than 7,000 marine species, including green sea turtles, endangered ducks, Hawaiian monk seals, seabirds, and coral reefs, as well as providing feeding grounds for whales.

Another large protected area, around the Chagos Archipelago in the Indian ocean, forms a 210,000-square-mile zone. While the status of these islands is a matter of international dispute, no one really argues that the sanctuary has been brilliantly effective at providing shelter for sharks, rays, tuna, and turtles. However, it is unfortunate that this is also being used to prevent the Chagosians from returning to their islands. By all accounts, the Chagosians have historically been and would make brilliant environmental stewards of their own oceans.

Unfortunately, such political and diplomatic arguments over boundaries, borders, and which country has the best claim to marine resources mean that vast natural areas are actively being blocked from protection. The 700,000-square-mile Antarctic preserve, for example, would have banned all fishing in the Weddell Sea to protect penguins, killer whales, leopard seals, and blue whales. Few areas in the world absorb more carbon dioxide than the waters around Antarctica, and protecting them would have damaged no meaningful economic activity. And yet some nations continue to play games by stopping it from being agreed in negotiations, content to use this amazing natural resource as a geopolitical bargaining chip.

Before we despair of political leaders making progress, there are lots of other ways we can help the ocean. Some new schemes are beginning to tackle the problem of agricultural run-off and the intense aquaculture in confined bays that has led to an overdose of nutrients degrading coastal ecosystems. Fish and shrimp farms are now growing kelp or mussels to help clean and filter the seawater. Such bio-mitigation has the benefit of producing multiple valuable products, while improving the overall vitality of the coastal area. Kelp is especially good at this because it is a primary producer. Recycling waste nutrients from aquaculture fish through kelp increases NPP while also removing these nutrients that otherwise could trigger harmful algal blooms.

Large-scale industrial aquaculture has grown to become a major protein and food source. Today more fish is farmed than is wild-caught, and the industry is rapidly modernizing to manage disease, reduce waste, and improve efficiency. Since most of the fish we like to eat are the carnivores at the top of the food chain, such as salmon and seabream, we used to feed them with wild-caught fish in the form of fish-meal and fish oil. But this so-called "fish-in-fish-out" problem, where wild-caught fish is used as a component of farmed-fish diets, is now being reduced.

Some unsavory practices continue, but in general fish farms are cleaning up the industry because it makes good business sense. Putting little fish called wrasses in with salmon keeps the cages clean, helps with oxygen exchange, and reduces parasites. Finding ways to vaccinate or stimulate the immune systems of fish is much more efficient and effective than pumping them full of expensive and counterproductive antibiotics. And there is now an increasing push to reduce the fishmeal content in aquaculture food—not least because it can cut costs. As aquaculture continues to mature, regulators are getting better at protecting the environment and ensuring animal welfare as well.

It says a lot about the state of our oceans that what we're now concerned about is wild-caught fish, which do not get the benefits of these controlled production practices and instead assimilate poisons from our polluted waters. It means that, counterintuitively, they are often less healthy than those that are farmed.

The richer humans become, the more protein—especially fish and poultry—they demand. Aquaculture can meet that demand with less environmental impact than other forms of meat. Farmed salmon, for instance, has a carbon footprint twenty times lower than that of beef. As cold-blooded creatures, fish do not need the quantity of food that warm-blooded cows, pigs, sheep, and goats require to heat and fuel their bodies. Fish can also be farmed in places like rivers, bays, and the open ocean which do not, like grazing pastures, use up agricultural land that is needed for crops. Because aquaculture is much more efficient at providing us with food than other forms of meat, it can make a major contribution to revitalizing our planet.

Can we improve photosynthesis?

The examples described in this chapter show at least some of the things we need to do as we try simultaneously to feed our population and keep our planet habitable. As science advances, new possibilities become available and we periodically get excited about futuristic ways of limiting the damage of climate change. Successful demonstrations have shown that giant chemical air scrubbers might help to clean carbon dioxide from the air. Inventors promote the idea of growing new foods in bioreactors. Since we humans are tinkerers, it is perhaps inevitable that we will even try to improve photosynthesis itself.

Why would we meddle with photosynthesis? When compared to the best photovoltaic cells, photosynthesis is simply inefficient. The average land plant converts 0.1 to 1 percent of the sunlight striking it into useful chemical energy. Crop plants, if well managed, achieve an overall rate of 1 to 2 percent of effective energy conversion. Even at peak performance, our own microalgae in our sun-filled desert ponds only operate at 6 percent efficiency. That is rather poor performance when compared to the latest six-junction photovoltaic cells with light concentrators that are nearing 50 percent of light to useful electric energy conversion, developed at the US National Renewable

Energy Laboratory (NREL). Clearly, there's a lot of potential for improving photosynthesis.

I am one of the scientists who have tried. Back in the mid-1980s at the Massachusetts Institute of Technology, I helped create the first site-directed mutants of something called Photosystem II.[83, 84] After the exhausted excitement of working several-hundred-hour weeks and sleeping in the lab, we discovered that it is difficult to improve on four billion years of evolution. The idea, however, was a good one. We were trying to understand how the photosynthetic machinery converts light into chemical energy.

Similarly, scientists have tried to tackle the problem of photorespiration. As discussed in Chapter 2, this is the paradox in which photosynthesizing organisms struggle to deal with the oxygen they produce, which reduces their ability to fix carbon by around 25 percent. There are multiple genetic engineering initiatives to reduce this photorespiration "waste" and increase the overall productivity of agricultural crops. Some plants, such as corn, have ways of storing carbon produced during the night so that they can enrich the photosynthetic conversion of carbon dioxide into complex chemicals during the day. There are projects currently underway that are seeking to introduce these carbon-concentrating mechanisms into other commercially significant crops. Scientists at the US Agricultural Research Service and the University of Illinois have, for instance, successfully increased photosynthetic efficiency in tobacco leaves by 20 percent.[85] Wild mustard, pumpkin, and green algal and bacterial genes were recombined and promoted, to optimize the photosynthetic metabolism. Now the team is seeking to import this upgraded photosynthetic system into crop plants, such as wheat, rice, or soy. A 20 percent increase in their productivity would mean that we could feed significantly more people without using more land.

Others have attempted to create "artificial leaves" or "artificial photosynthesis." These are important efforts to isolate, focus, and improve the advantages of photosynthesis in more efficient devices. Some create devices that perform the same function of capturing CO_2 and turning it into useful chemicals such as syngas (synthetic gas: a fuel-gas mixture), while others are using photovoltaic energy–driven membranes to capture CO_2 to concentrate it with electrically charged membranes. In recent years, a major step toward artificial, but still biological, photosynthesis has been achieved by isolating chloroplast components in microscopic assemblages and separating them from the rest of the cell to boost their efficiency. The Franco-German team behind this project envisions that their enhanced photosynthetic machinery might be used to synthesize, for example, complex pharmaceutical molecules with the power of light.

All these inventive approaches have advantages and limitations. Often the net environmental impact is uncertain or very costly. In some cases, these improvement efforts simply transfer the efficiency or cost problem elsewhere. Most of these new devices lack long-term sustainability; they also take a lot of time and large amounts of money. The latest photovoltaic cells took more than fifty years to reach their current peak efficiencies, while the average installed base of photovoltaic cells linger between 15 and 20 percent efficiency.

Yet we need to try every advance, assess every potential solution, use every method to engage people and keep them engaged. Growing our world again will require hyperbolic entrepreneurism, new technology, and traditional methods. As we have seen, science is helping us in all sorts of ways—upgrading pasturelands through managed grazing, inventing low-energy methods for making ammonia, providing new methods of recycling organic materials. There are nearly 63,000 power plants, hundreds of millions of hearths, and billions of cars, all burning fossil fuels. It requires as many ideas, methods, and approaches to balance their legacy.

Of course, we need to improve agriculture and photosynthesis to use the agricultural land we have more effectively to grow food for all of us. We must continue to innovate, experiment, and push the limits of what is possible in terms of carbon removal and to increase the efficiency of photosynthesis itself, even if this is difficult. But that misses the opportunity to use photosynthesis in its current state, as it is today.

Photosynthesizers grow in every environment: under the ice in Antarctica; stuck to the side of granite buildings; in forests, fields, oceans, and modern urban greenhouses. Even if less efficient, plant life covers a much larger surface area than all other carbon-capturing technologies combined, including both natural processes and artificial ones—the latter taking years to design, build, and commission. Plants have "installed" themselves already; they capture carbon while repairing, seeding, and multiplying themselves, feeding all ecosystems, and flexibly responding to environmental change, seasonality, and weather. Nobody must come and clear leaves of dust in the same way we need to clear photovoltaic cells to keep them operating. Most importantly, photosynthetic organisms are free to use today, and we have known, for thousands of years, how to propagate them. All societies know how to grow plants in their local environment, instead of waiting for a future technological breakthrough in efficiency.

Instead of schemes to shade the sun, fill the atmosphere with aerosols, or blast us to other planets, we have the planet-transforming potential of plants working for us today. I am an inventor and I am excited by technology, and I also value prosaic projects that enrich us biologically and economically—like making a bit more dirt and planting a lot more trees. We can transform our planet for the better, with self-sufficient—if not the most efficient—photosynthetic technology right now.

CHAPTER 7

HARVEST THE LIGHT

Deep down, most of us already know the truth: No matter what our class, creed, or country, we are all participating in the devastation of our environment. A rich heiress flying off to a ski resort in a private jet may think she has little in common with a low-paid laborer building a road through African forest. But both of them are extracting more natural resources than they are replacing. Whether we are cooling offices in steaming Asian cities with air-conditioners or overheating outposts in the Arctic to mine oil and gas, we are all voraciously consuming the air, sea, and land of our only home. You and I, we are the locust swarm that is eating the planet and wearing down its resilience. News of environmental catastrophe that was once the exception—whether hurricanes in America, fires in Australia, floods in

Bangladesh—is now becoming the norm. Even the COVID-19 pandemic almost certainly had its origins in deforestation and agricultural development, as these have forced wild animals closer to humans, allowing new strains of infectious diseases to cross over from one species to another.

Amid these calamities, as with so many other issues, political leaders cast blame and industrial lobby groups sow confusion to justify environmental exploitation. Responsibility does not lie with some external force like a solar flare or gigantic volcano eruption, any more than it does with one nation, president, corporation, or ethnic group. It is human beings—all of us—who throughout history have unleashed plague and pestilence, not to mention famine, firestorms, and floods, on our Earth.

The first step to solving a problem is to recognize it. And the great beauty of residing on a planet with the capacity for photosynthesis is that we do not live in a finite, zero-sum world but one filled with growth and endless possibility. These are also *our* possibilities. From the first scientists realizing that plants were not eating the soil, to Norman Borlaug finding ways to feed billions, we have learned that the advance of our species does not have to mean the destruction of others. Instead, we are discovering again and again how Earth's multitrophic, interwoven, planet-spanning biological systems are connected. Humans are still a product of the very natural environment that we are exploiting. We are integral to the system. We use the services of the ecosystem with every breath we take and provide ecosystem services with every breath we exhale. We feed the strangest microbiology living in us and on us—from the array of life in our guts or our eyelash follicles, to the cosmopolitan community living in the microbiome of our navels.

Exciting new areas of research into ecosystem dynamics and collaboration across completely different organisms are opening up. We still have much to discover about how plants communicate with each other and how they influence us. Every advance we make brings better understanding and a deeper appreciation of how we can actively collaborate to make more nature because, whether we deserve to be or not, we are now the stewards of our planet.

With responsibility comes opportunity, not least for extensive new job creation. Nature keeps creating more resources, more opportunities, and new methods of working. In the midst of all the despair about our future, photosynthesis should be the light that gives us hope. Weird-colored bacteria can green our deserts. Algae can decontaminate poisoned soil. Maybe photovoltaic wasps, photosynthetic snails, and endosymbiotic salamander algae are pointing the way to a future where we incorporate photosynthesis in more living things as an active collaboration to save our shared environment.

The problem is that all too often we are still afraid of—and in a struggle for survival with—nature. In his poem, *In Memoriam*,[86] Alfred, Lord Tennyson conjures a world of ferocious beasts hunting helpless prey. But he also recognized the difference between the plight of an individual and the powerful constancy of nature's support for us as a community.

> *Are God and Nature then at strife,*
> *That Nature lends such evil dreams?*
> *So careful of the type she seems,*
> *So careless of the single life . . .*

It is this leap, from fearing only for our "single life," to caring for others, our "type," that should give us confidence. We must reconnect with and care for the wild expanse and wealth of nature in us and around us.

We already know our mental health is better if we spend two hours a day outdoors in parks or countryside. We declare our love in terms of flowers and fruity poems. We describe our paradises as green and pleasant lands, Elysian Fields, or the Garden of Eden. These associations have survived thousands of years of human domestication and civilization. We can do far more to embrace the biology within and without. The flavors, colors, and smells of nature are a reminder that life is not only about our economic value and that it is time to recognize the beautiful, bountiful force that has shaped us just as much as we are shaping it now.

Our satellites looking down from space have helped by giving us a truly global perspective. Our microscopes looking inward are discovering complex ecosystems of microorganisms living within and on us. Our technology looking through deep mines of data can track the informational content of systems that mean that a drought in Russia affects food prices in Mexico. We can now see our place in a greater scheme where all actions are linked. We are filled with plastics that we threw away and that found their way straight back into our food. We need to stop pretending it is someone else's problem.

Perhaps one day we will live in a world where global business tycoons measure their worth by the size of their sequoia trees, where entrepreneurs compete to grow the best mangrove swamp, and young hearts are wooed by gifts of painted snail shells grown in managed forests.

The simple and urgent question facing us all is whether we can change our mindset so that we stop being the problem and become the solution. There is a fundamental obstacle to this: Our laws, economic systems, and culture do not value our planet's natural vitality enough.

Economists recognize that the prices we pay for services and goods in our daily life rarely reflect their true worth. All kinds of subsidies, taxes, perverse incentives, and profiteering distort the market. For example, water in Cairo should be a scarce resource in a desert country. Yet it is taken from the Nile and the distribution of it is paid for by the government. As a result, few people have a reason to invest in new infrastructure, irrigation systems, or even to mend a leaky toilet.

When costs are paid for by someone else, they are said to have been externalized. And, all across the world, we are externalizing the costs of massive damage to our ecosystems by foisting them on taxpayers, other countries, and especially the next generation. Much of the historic wealth we have created is based on the direct extraction of natural resources and making others pay for clearing up the mess.

The previous chapter discussed how the Stern report back in 2006 sought to estimate the cost of our continued failure to act against carbon pollution. Since then, there have been many carbon-pricing experiments and some faltering progress. But governments are continuing to allow or enable the distortion of markets in support of business interests, to keep groups of voters happy, or to compete with other countries. The result is that it still makes twisted sense to pollute the world.

The challenge for governments now is to find a fair balance between taxes designed to make us understand the real price of our current behavior and incentives to encourage an increase in Net Primary Productivity. Few people like paying taxes but most of us understand the need for some form of levy so that societies can function and sustain justice systems, schools, and sewerage systems. Conversely, we like incentives that reward good behavior. Governments can impose new taxes that discourage emissions, fossil carbon content, inefficiency, one-way packaging, waste, or the energy content of products. They can make shoppers pay a modest tax on plastic bags or impose new costs on motorists transporting themselves around in bloated and mostly empty vehicles. They can also try to change behavior by offering tax breaks to farmers to become better stewards of the land and subsidies to people who plant trees or improve our soil.

It should be blindingly obvious that one of the biggest problems facing our planet is that governments are not, by and large, doing this enough. Most governments have token environmental programs that just about paper over some of the worst destruction. Some are better than others. Sadly, the reality is that most of our governments use our taxes to finance the destruction of the environment, for archaic and unimaginative short-term gains. An array of choices about different taxes and incentives—carrots and sticks—are put in front of us at elections where we can bring change through voting. More often than not, voters are persuaded they have no choice. We are told that only carbon emissions guarantee "economic stability," that there is a trade-off between jobs and environmental preservation, or that international regulation has to be the prerequisite before we do anything. And we must start doing so now. The war in Ukraine has shown that our dependence on cheap fossil fuels makes us beholden to wild political manipulations. Instead of scrambling to find alternative cheap fossil fuels, we need to build more renewable energy capacity and more self-reliance. It will make us environmentally, economically, and politically safer than importing yesterday's ideologies together with decayed old carbon.

Similar arguments were made by the politicians who shrugged their shoulders when famine struck Ireland or by sugar producers to justify slavery. We do not have to accept these arguments. We have changed the rules before, and we can do so again. If we live in a democracy it is up to us to vote for representatives that will reverse destructive policies like crazy fossil fuel subsidies, which continue to expand coal mining and oil drilling. We can elect governments that will redirect this money into financing truly renewable energies so that our children are not forced to live on a steamy grey planet. We can blame politicians as much as we like but, in the end, it is people who elect or enable them. We must demand better.

Our chances of winning this battle for change will improve only if people are armed with better information. Any action to change behavior requires data because it informs rational decisions about what is wrong and what can be done to put it right. We need to be measuring carbon content, water footprint, land use, and waste. Measurement enables policy makers to quantify the task in hand and governments to act fairly. Publicly available data help voters to hold governments to account, and prompts citizens to take legal action and international authorities to enforce the rules.

Better data also help people make better choices for themselves and perhaps to trade off priorities. Someone flying a short-haul two-hundred-mile shuttle flight from New York to Boston for a business meeting emits roughly 250 pounds of carbon dioxide, even in an economy seat. Whether that same person then recycles a paper coffee cup at the end of that flight really does not matter in comparison. That might seem obvious, but people are still bombarded with propaganda telling them that untaxed aviation fuel is an international issue that has nothing to do with domestic carbon emissions. We live in a fog of thousands of market distortions, misaligned incentives and wilful wrong-headedness.

Perhaps more and better efforts should be made to communicate these facts effectively. Dramatic advertising, including those disgusting pictures of tumors on cigarette packs, has been used to cut smoking. Why is there no such campaign against food loss and waste? This would go such a long way to ensuring we can feed the world's population.[87] Instead, most of our advertising industry is devoted to telling us to eat more and consume more. It is actually possible to go jogging without a bladder-equipped backpack containing a sugary electrolyte solution or wearing sweat-wicking artificial fibers and a GPS system with a heart monitor.

Do we really need a public information campaign before we improve our behavior? The urgency of the issue was underlined by a recent project in which 223 researchers joined forces to physically measure half a million trees in 813 forests. They found that the edges of recently logged areas warm by more than 3.6°F (2°C), and that is enough to dry the forest and prevent seeds from spreading.[88] Between logging, roads, fires and shifting weather patterns, these essential rain forests are straining under the unrelenting assault to make room for cow pastures and soya fields. Should we ignore our own responsibility while politicians bicker over a binding international agreement on carbon taxation?

Of course not.

We can be the solution

Too often, we feel overwhelmed by the continuous stream of bad news about forest fires, storms, and disease. Too many of us are told, in the face of powerful interests, that we have no power. Too few of us try to reconcile our conflicting desires for safety, food, and pleasure without making excuses.

Sure, there is no magic bullet. And yet, just as we have all contributed to the problem, we can all contribute to the solution. The vast improvements we have made in global health and wealth over the past two hundred years show that seemingly insurmountable obstacles can be overcome.

In 1798, the English cleric Thomas Robert Malthus anonymously published a book called *An Essay in the Principle of Population*.[89] It was enormously influential, leading to a nationwide census and serious debates about birth control. His warning was that "the power of population is so superior to the power of the earth to produce subsistence for man, that premature death must in some shape or other visit the human race." Such predictions of catastrophe have resurfaced regularly through the generations. In 1968, the biologist Paul R. Ehrlich wrote in his bestseller *The Population Bomb* that "the battle to feed all of humanity is over . . . In the 1970s and 1980s hundreds of millions of people will starve to death."[90]

But since Malthus, our population has doubled four times in a row, and ordinary people are living longer and better than ever. Yes, in that time, we have increased the exploitation of the environment in countless stupid ways. And yet we have also educated ourselves. Half a century on from Ehrlich, the destruction of our planet's vitality is worse than ever, but with a constructive outlook we have still, just about, found a way of feeding the world.

In this book, I have tried to balance my anxiety about the loss of nature with a recognition of just how brilliant humans are at achieving extraordinary things. Hans Rosling, the Swedish academic, has captured some of this in his amazing videos of countries, represented by bubbles, being lifted out of poverty.[91] The World Values Survey[92] similarly demonstrates the extent to which we humans have changed in terms of how we think and feel since the 1980s. Even if there are serious, regular, and disastrous setbacks, societies are generally becoming more rational than it might seem. Think about it: We can track how societies modernize and change their thinking in our own lifetimes. Unlike the painfully slow scientific progress set out in the opening chapter of this book, our awareness of the scale of the challenges and our energy in putting serious thought toward potential solutions have rapidly gathered pace.

A few years ago, we were expecting the global population would grow to at least fourteen billion people. Today, as mentioned earlier, the Population Division of the Department of Economic and Social Affairs of the UN Secretariat forecasts we will top out at 10.88 billion people by the end of this century. Such a reduction, of more than 3.1 billion,[93] has been achieved by a range of social improvements, including rising affluence in developing countries and the provision of basic health care and antenatal care to the world's poorest people. When children can be protected from preventable deaths and when women are not in mortal danger every time they go into labor, the rate of childbirths dramatically decreases. These are not enormously expensive interventions. In 1980, only 20 percent of the world population was vaccinated against a smattering of diseases. Today, we have reached between 80 and 85 percent coverage against a much broader spectrum of diseases. Few things have had as large an impact on human reproduction as educating women. In just about every society across every income bracket, educated women have fewer children than those denied such opportunities. Alternative estimates suggest the population rise may be even smaller, reaching just 8.79 billion by 2100 if the sustainable development goals on woman's education and contraception are met.[94]

Even with these slower growth rates, our population is still set to rise for multiple generations until we reach our peak. Old generations, like mine, will die. Billions of new people will have to be taught and inspired. Environmentally conscious young minds represent a huge responsibility for educators and a brilliant opportunity for reconnecting our emotions with the vitality of our planet.

All of us can start making changes, big and small, both for ourselves and by informing others. We can help change the world whether we are young or old, rich or poor, living in the heart of a metropolis or in a shack at the edge of the world. It can be done while we take our kids to school. It can be done on our lunch break from our exhausting job. It can be done when we remember to turn the light out before we go to sleep. It can be done right now, today.

We have all the tools we need. We can all grow more photosynthetic biomass. We can do it anywhere, anytime, anyhow.

For instance, I grew my own wedding jacket out of sturdy, bright-green rye grass. My father-in-law suspected something was going on, so, in a misguided effort to make it a surprise, I nurtured and watered the jacket on a neighboring grass embankment. Unfortunately, on the morning of my wedding it was pouring with rain and my waterlogged green outfit was dripping water down my legs into a puddle during the service. In stark contrast to the beautiful white dress worn by my wife, I looked like a giant green soggy caterpillar. Nor did it increase primary productivity that much. There are undoubtedly better and more sustainable schemes out there.

Very few of us live on a farm, in a forest, or have enough land to grow anything of note. Wherever we live, though, we are still flooded with abundant life energy every day. And we need to harvest that light by planting trees—or at least voting for local government and urban planners who will prioritize the planting and protection of trees, mangroves, or heather, for example. There are urban parks, rooftops, hedges, and edges that we can plant with trees, shrubs, and flowers. We can plant more along the sides of highways, agricultural fields, rewilded pastures, and industrial sites.

Nor do we necessarily have to just grow trees. As outlined in the previous chapter, there are lots of other ways to increase biomass. Make bird boxes, build beehives, take care of community gardens, or make clever pots and beds that can fit into urban spaces. Compost and make fertilizer from organic waste. There is so much organic waste that can be shredded and placed in alternating layers of green and woody biomass to make fresh, rich soil within weeks. What matters is to grow something that is enduring, that captures carbon and is adapted to survive in the local environment. We all can learn to live with our photosynthetic partners again. There are countless charities and helpful organizations that will help us to regrow the world.

Even we city dwellers trapped in tiny apartments can grow a lot of things. It costs virtually nothing. As a penny-pinching first-time employee, I grew garlic, pineapple, and pretty much anything else I could find at the supermarket on the windowsill of my one-room flat. With a little patience, you can use planters that require less than a minute per day to take care of. It is also calming and personally rewarding, and potatoes and peas produce really pretty flowers.

Most fruit has live seeds in it. Avocado seeds germinate slowly but the process can be accelerated by peeling away the outer layer of the seed and adding aspirin to the water to prevent mold. If growing things from seeds is too difficult, try using cuttings. For example, it is possible to grow multiple saplings by taking 6- to 10-inch cuttings from a Christmas tree and sticking them in water with rooting hormone. They can then be grown in rock wool or compost. By the next Christmas, there will be dozens of clones from last year's tree.

Another method, called air-layering, enables you to create root stock, mid-branch, on large plants like passion fruit, figs, apple, avocado, and citrus, in just two to eight weeks without sacrificing the original tree. All of these techniques can be found online, and there are countless instructional videos on YouTube. Using mainly common household materials, you will find ingenious solutions to make this work in your context.

My friend balances his passion for the environment with his love for beer. Every time he empties a beer can, he scoops in a handful of soil and plants a sunflower seed. He lives in a very modest room surrounded by a forest of sunflowers that cover every surface—including windowsills and his rooftop—and spill over into his hallway. Munching on his home-made, roasted sunflower seeds and drinking his beer, he is still making a difference.

I prefer growing seedlings in little non-woven fabric seed bags that protect them from intertwining their roots so that I can take one out without disturbing the others. Currently, my home is sprouting cherry, plum, prune, hazelnut, coffee, fuchsia, passionfruit, red date, annatto, and Canadian serviceberry seeds. When the saplings get too big for the little seed bags, I cut the fabric away and plant them in flowerpots. By the time this book is published, I hope to have hundreds of sprouting saplings to give away. All I needed was less than 3 square feet of space.

As a child, I grew slime-molds and fungi under my bed. It was years before my mother found out, and I eventually had to put them in the hallway in front of our apartment. Now, on our balcony, we are raising many trees in pots. We have already planted a tree for every family member. Seven of them are apple trees from a rare old variety that is no longer cultivated commercially. The apples are delicious, and they have sentimental value for my family because they remind my children of their grandmother. We have grown acorns from the ancient English oak under which young Queen Elizabeth, five hundred years ago, found out that she was not going to be executed. Today the oak is a thousand years old and still making seeds, and we have a scion of it on our terrace. And, oh yes, we have two very large banana plants in ridiculously tiny pots, courtesy of my son.

In the smallest of ways, watching all these tiny seeds germinate provides an antidote to despair. Each little germ fulfills its tiny promise, that something new, constructive, and enduring is emerging. We can confidently predict that many if not most of these sprouts will survive, and my sons are excellent at planning ahead, anticipating the next season, when their plants burst forth with more leaves. Things go wrong. Snails eat young sprouts; saplings dry up or drown in a rainstorm and overzealous mums throw out unsightly little shrubs. But with a little persistence and a lot of help from YouTube, all of us can learn how to germinate, propagate, clone, and grow just about anything.

These small interventions in the global carbon cycle and in planetary productivity will actively improve the future. There are nearly eight billion of us. If we all pitch in, we will have a powerful impact. Of course, there is a key role for governments, scientists, and international agencies in all this, but we can also take direct, personal, and very rewarding action ourselves. There is nearly infinite power to drive the process of plant growth. It is available every day. In each seed is millions of years of the most sophisticated programming, with the most amazing protective infrastructure and the latest in disease resistance. And, when they flourish, they are beautiful and help all of us to change the way we speak, think, and feel about the natural world.

This book has tried to show not just the magic of photosynthesis, but its resilience—and how our understanding of it has advanced despite the Spanish Inquisition, ridicule, misunderstanding, political turmoil, and intellectual upheavals. Although many of these discoveries were driven by the need to grow more food and prevent starvation, the brave and quirky scientists who made them also wanted to understand the beauty of these organisms, how they made—then remade—the world.

It has been through photosynthesis that the total amount of life on our planet has generally increased, until recently when humans came along and began extracting natural resources in a way that now threatens to make much of the world less productive.

But humans have also shown they have risen to every challenge. We know there are highly productive places like mangroves, coral reefs, and rain forests that should be protected because they absorb and sequester more carbon than anywhere else, while providing food and shelter for a vast variety of animals, including ourselves. We also know that agricultural productivity can be increased, and its carbon footprint decreased, so that it doesn't encroach further on Earth's natural vitality.

There are many changes in laws, taxes, and international agreements that can help make humans better stewards of our shared environment and increase NPP, while also creating new jobs and business opportunities. We can improve grazing, replant forests and revitalize derelict savannas. Cities can be greened, and oceans nurtured, without our having to create fantastical new or expensive technologies.

And it is relatively easy to change our own behavior and have real impact by making more life for everybody. We should do more to enjoy the seductive virtues, colors, spices, and smells of plants. Planting seeds is neither quaint nor small scale. It is about participating in the biggest, longest-lasting, most successful and enriching growth program ever.

Photosynthesis was fine-tuned by billions of years of innovation and experimentation; it is the most powerful force that exists for the restoration of our world; and it is freely available to us all and fundamental to our identity as humans. On every continent, in every society, no matter what our profession, we already know how to promote it.

Harvest the light and make our planet grow again.

NOTES

1. Bodenheimer, F. S., *The History of Biology: An Introduction*. W. Dawson and Sons, 1958.

2. Howe, H. M., A root of van Helmont's tree. *Isis*, 1965, Vol. 56, pp. 408–419.

3. Helmont, J. B. v., *Ortus Medicinae, vel opera et opuscula omnia*. Franciscus Mercurius van Helmont, 1648.

4. Mariotte, E., *Discours de la nature de l'air, de la végétation des plantes. Nouvelle découverte touchant la vue*. Gauthier-Villars, 1923.

5. Hales. S., *Vegetable staticks, or, An account of some statical experiments on the sap in vegetables: being an essay towards a natural history of vegetation: Also, a specimen of an attempt to analyse the air, by a great variety of chymio-statical experiments*. W. & J. Innys & T. Woodward, 1727.

6. Darwin, E., *The Botanic Garden, A Poem in Two Parts. Part 1 The Economy of Vegetation. Part 2. The Loves of the Plants*. J. Johnson, 1791.

7. Liebig, J.F. v., Priesner. C., *Neue Deutsche Biographie*, 1985, Vol. 14, pp. 497–501.

8. Barnes, C. R., On the food of green plants. *Botanical Gazette*, 1983. Vol. 18, pp. 403–11.

9. Arrhenius, S., Über den einfluss des atmosphärischen kohlensäurengehalts auf die temperatur der erdoberfläche. 1896, *Proceedings of the Royal Swedish Academy of Science*, Vol. 22, pp. 1–101.

10. Roser, E., & Ortiz-Ospina, M., World population growth, 2017. https://ourworldindata.org/world-population-growth.

11. Blankenship, R., *Molecular Mechanisms of Photosynthesis*. Wiley, 2002.

12. Barker, H. A., & Hungate, R. E., *Cornelius Bernardus van Niel, 1897–1985, A Biographical Memoir*. National Academy of Sciences, 1990.

13. Sagan, C., & Mullen, G., Earth and Mars: Evolution of atmospheres and surface temperatures. *Science*, 1972, Vol. 177, pp. 52–56.

14. Leman, L., Orgel, L., & Ghadiri, M. R., Carbonyl sulfide-mediated prebiotic formation of peptides. *Science*, 2004, Vol. 306, pp. 283–86.

15. Bell, E. A., Boehnke, P., Harrison, T. M., & Mao, W. L., Potentially biogenic carbon preserved in a 4.1 billion-year-old zircon. *Proceedings of the National Academy of Sciences*, 2015, Vol. 112, pp. 14518–21.

16. Dodd, M. S., Papineau, D., Grenne, T., et al., Evidence for early life in Earth's oldest hydrothermal vent precipitates. *Nature*, 2017, Vol. 543, pp. 60–64.

17. Glansdorff, N., Xu, Y., & Labedan, B., The last common universal ancestor. *Biology Direct*, 2008, Vol. 3, p. 29.

18. Holland, H .D., The oxygenation of the atmosphere and oceans. *Philosophical Transactions of the Royal Society*, 2006, Vol. 361, pp. 903–15.

19. Warneck, P., *Chemistry of the Natural Atmosphere*. Academic Press, 2000.

20. Haviga, J. R., & Hamilton, T. L., Snow algae drive productivity and weathering at volcanic rock-hosted glaciers. *Geochimica et Cosmochimica Acta*, 2019, Vol. 247, pp. 220–42.

21. Weule, G., News. ABC Net, Twelve of the oldest fossils we've discovered so far, 2016. https://www.abc.net.au/news/science/2016-09-01/twelve-of-earths-oldest-known-fossils/7796176.

22. Cary, S. C., McDonald, I. R., Barrett, J. E., & Cowan, D. A., On the rocks: the microbiology of Antarctic Dry Valley soils. *Nature Reviews Microbiology*, 2010, Vol. 8, pp. 129–38.

23. Clarkson, M. O., Kasemann, S. A., Wood, R. A., et al., Ocean acidification and the Permo-Triassic mass extinction. *Science*, 2015, Vol. 348, pp. 229–32.

24. Smil, V., Detonator of the population explosion. *Nature*, 1999, Vol. 400, p. 415.

25. Lewin, R. A., Prochloron. *Phycologia*, 1975, Vol. 14, pp. 153–60.

26. Brockmann, J. H., & Lipinski, A., Bacteriochlorophyll g. A new bacteriochlorophyll from *Heliobacterium chlorum*. *Archives of Microbiology*, 1983, Vol. 136, pp. 17–19.

27. Noriaki, K., Yoshimasa, K., & Tatsuo, T., *Acidobacterium capsulatum* gen. nov., sp: An acidophilic chemoorganotrophic bacterium containing menaquinone from acidic mineral environment. *Current Microbiology*, 1991, Vol. 22, pp. 1–7.

28. Brenner, D. J., Krieg, N. R. & James, T. S., "Introductory Essays." In Garrity, George M. (ed.). *Bergey's Manual of Systematic Bacteriology*. (2nd ed.) Springer, 2004, (orig-pub: Williams & Wilkins, 1984), p. 304.

29. Lami, R., Cottrell, M. T., Ras, J. et al., High abundances of aerobic anoxygenic photosynthetic bacteria in the South Pacific Ocean. *Applied and Environmental Microbiology*, 2007, Vol. 73, pp. 4198–205.

30. Beatty, J. T., Overmann, J., Lince, M. T., et al., An obligately photosynthetic bacterial anaerobe from a deep-sea hydrothermal vent. *Proceedings of the National Academy of Science*, 2005, Vol. 102, pp. 9306–10.

31. Schimper, A. F. W., Über die entwicklung der chlorophyllkörner und farbkörpe. *Botanische Zeitung*, 1883, Vol. 41, pp. 105–14, 121–31, 137–46, 153–62.

32. Mereschkowski, K., Über natur und ursprung der chromatophoren im pflanzenreiche. *Biologisches Centralblatt*, 1905, Vol. 25, pp. 593–604.

33. Wallin, I. E., On the nature of mitochondria. I. Observations on mitochondria staining methods applied to bacteria. II. Reactions of bacteria to chemical treatment. *American Journal of Anatomy*, 1922, Vol. 30, pp. 203–29.

34. Wallin, I. E., *Symbionticism and the origin of species*. Williams & Wilkins Company, 1927.

35. Teresi, D., Discover interview: Lynn Margulis says she's not controversial, she's right. *Discover* magazine, 2011.

36. Pollan, M., *The Botany of Desire*. Random House, 2001.

37. Burkholder, J. M., Noga, E. J., Hobbs, C. W., et al., New "phantom" dinoflagellate is the causative agent of major estuarine fish kills. *Nature*, 1992, Vol. 358, pp. 407–10.

38. IFLS, Marine plankton found on surface of International Space Station, 2014. https://www.iflscience.com/space/marine-plankton-found-surface-international-space-station.

39. Valmalette, J. C., Dombrovsky, A., Brat, P., et al., Light-induced electron transfer and ATP synthesis in a carotene synthesising insect. *Scientific Reports*, 2012, Vol. 2.

40. Kerney R., Kim E, Hangarter R. P., et al., Intracellular invasion of green algae in a salamander host. *Proceedings of the National Academies of Sciences*, 2011, Vol. 108, pp. 6497–502.

41. NASA Earth Observations (NEO), Net Primary Productivity, 2000–2016. https://earthobservatory.nasa.gov/global-maps/MOD17A2_M_PSN.

42. Field, C. B., Behrenfeld, M. J., Randerson, J. T., & Falkowski, P., Primary production of the biosphere: integrating terrestrial and oceanic components. *Science*, 1998, Vol. 281, pp. 237–40.

43. Buesseler, K. O., Boyd, P. W., Black, E. E., & Siegel, D. A., Metrics that matter for assessing the ocean biological carbon pump. *Proceedings of the National Academy of Sciences*, 2020, Perspective.

44. Bar-On, Y. M., Phillips, R., & Milo, R., The biomass distribution on Earth. *Proceedings of the National Academies of Sciences*, 2018, Vol. 115, pp. 16506–11.

45. Burgess, M. G., & Gaines, S. D., The scale of life and its lessons for humanity. *Proceedings of the National Academies of Sciences*, 2018, Vol. 115, pp. 6328–30.

46. Hatton, I. A., McCann K. S., Fryxell J. M., et al., The predator-prey power law: Biomass scaling across terrestrial and aquatic biomes. *Science*, 2015, Vol. 349.

47. Trebilco, R., Baum, J. K., Salomon, A. K., et al., Ecosystem ecology: size-based constraints on the pyramids of life. *Trends in Ecology & Evolution*, 2013, Vol. 28, pp. 423–31.

48. Beerling, D. J., & Berner R. A., Impact of a permo-carboniferous high O_2 event on the terrestrial carbon cycle. *Proceedings of the National Academies of Sciences*, 2000, Vol. 97, pp. 12428–32.

49. Andrei, V., Reuillard, B., & Reisner E., Bias-free solar syngas production by integrating a molecular cobalt catalyst with perovskite–BiVO4 tandems. *Nature Materials*, 2020, Vol 19, pp. 189–94.

50. Revision of World Population Prospects. UN DESA Population Division, 2019. https://population.un.org/wpp.

51. Rosling, H., *Factfulness*, Sceptre Books, 2018.

52. Bar-On, Y. M., Phillips, R., & Milo, R., The biomass distribution on Earth. *Proceedings of the National Academies of Sciences*, 2018, Vol. 115, pp. 16506–11.

53. Stiling, P., Ecology: *Theories and Applications*. Prentice Hall, 1996.

54. Paul, J., Hanson, Griffiths, N. A., Iversen C. M., et al., Rapid net carbon loss from a whole-ecosystem warmed peatland. *AGU Advances*, 2020, https://doi.org/10.1029/2020AV000163.

55. NPR.org, Your grandparents spent more of their money on food than you do, 2015. https://www.npr.org/sections/thesalt/2015/03/02/389578089/your-grandparents-spent-more-of-their-money-on-food-than-you-do?t=1561615587117.

56. OECD-FAO Agricultural Outlook, 2012. https://doi.org/10.1787/19991142.

57. Wallace, R. J., Sasson G., Garnsworthy, P. C et al., A heritable subset of the core rumen microbiome dictates dairy cow productivity and emissions. *Science Advances*, 2019, DOI: 10.1126/sciadv.aav8391.

58. Coady, D., Parry, I., Sears, L., & Shang, B., How large are global energy subsidies? IMF Working Papers WP/15/105, 2015.

59. Ibid.

60. Haberl, H., Erb, K. H., Krausmann, F., et al., Quantifying and mapping the human appropriation of net primary production in earth's terrestrial ecosystems. *Proceedings of the National Academies of Sciences*, 2007, Vol. 104.

61. Pharo, P., Oppenheim, J., Pinfield, M., et al., Growing better: ten critical transitions to transform food and land use. *The Food and Land Use Coalition*, 2019.

62. The Stern Review on the Economic Effects of Climate Change. Population and Development Review, 2006. https://doi.org/10.1111/j.1728-4457.2006.00153.x.

63. Stern, N. *The Observer*, Nicholas Stern: cost of global warming "is worse than I feared," 2016. https://www.theguardian.com/environment/2016/nov/06/nicholas-stern-climate-change-review-10-years-on-interview-decisive-years-humanity.

64. World Bank, Carbon pricing dashboard, 2020. https://carbonpricingdashboard.worldbank.org/map_data.

65. Cost the Earth sources. BBC Earth, 2015. http://www.bbc.com/earth/story/cost-the-earth-sources.

66. Grooten, M., & Almond, R. E. A., *Living Planet Report—2018: Aiming Higher*, WWF, 2018.

67. Cost the Earth sources. BBC Earth, 2015. http://www.bbc.com/earth/story/cost-the-earth-sources.

68. The Greatest Story Never Told, Iowa Public Radio, 2017.

69. Watts, J., *The Guardian*. Third of Earth's soil is acutely degraded due to agriculture, 2017. https://www.theguardian.com/environment/2017/sep/12/third-of-earths-soil-acutely-degraded-due-to-agriculture-study.

70. Dickinson, M., & Mace, G., Climate change and challenges for conservation. Grantham Briefing Paper 13, 2015. https://www.imperial.ac.uk/grantham/publications/briefing-papers/climate-change-and-challenges-for-conservation---grantham-briefing-paper-13.php.

71. Eakin, H., Bojórquez-Tapia, L., Janssen, M. A., et al. Urban resilience efforts must consider social and political forces, *Proceedings of the National Academies of Sciences*, 2017, Vol. 114.

72. OECD Report, Joint Working Party on Agriculture and the Environment, Improving energy efficiency in the agro-aood chain, 2016. https://www.oecd.org/officialdocuments/publicdisplaydocumentp df/?cote=COM/TAD/CA/ENV/EPOC(2016)19/FINAL&docLanguage=En.

73. Monforti-Ferrario, F., Dallemand J. F., Pinedo Pascua I., et al, Energy use in the EU food sector: State of play and opportunities for improvement. *JRC Science and Policy Report*, 2015.

74. USGS, Nitrogen Statistics and Information. https://www.usgs.gov/centers/nmic/nitrogen-statistics-and-information.

75. Phys.org, Novel reaction could spark alternate approach to ammonia production, 2018. https://phys.org/news/2018-05-reaction-alternate-approach-ammonia-production.html.

76. Phys.org, Researchers dramatically clean up ammonia production and cut costs, 2019. https://phys.org/news/2019-04-ammonia-production.html.

77. Gao, W., Guo, J., Wang P., et al., Production of ammonia via a chemical looping process based on metal imides as nitrogen carriers, *Nature Energy*, 2018, Vol. 3, pp. 1067–75.

78. McLaughlin, L., Forest Green Rovers granted planning permission for all-wooden stadium. *The Guardian*, 2019. https://www.theguardian.com/football/2019/dec/29/forest-green-rovers-granted-planning-permission-for-all-wooden-stadium.

79. Galina, C., Organsch, A., Reyer, C. P. O., et al., Buildings as a global carbon sink. *Nature Sustainability*, 2020, Vol. 3, pp 269–76.

80. Crowther, T. W., Glick H. B., Covey K. R., et al., Mapping tree density at a global scale, *Nature*, 2015, Vol. 525, pp. 201–5.

81. Bastin, J-F., Finegold, Y., Garcia, C., et al., The global tree restoration potential, *Science*, 2019, Vol. 365, pp. 76–79.

82. Makarieva, A., & Gorshkov, V., The Biotic Pump: condensation, atmospheric dynamics and climate, *International Journal of Water*, 2010, Vol. 5, pp. 365–385.

83. Jovine, R. V. M., Bylina, E. J., Tiede, D. M., & Youvan, D. C., Mutations directed at Tyrosine L162 of the Bacterial Photosynthetic Reaction Center. [book auth.] Lamb, C., & Beachy, R., Wiley-Liss, *New Series* Vol. 129, 1990.

84. Bylina, E. J., Jovine, R. V. M., & Youvan, D. C., A genetic system for rapidly assessing herbicide that compete for the quinone binding site of Photosynthetic Reaction Centers. *Bio/Technology*, 1989, pp. 69–74.

85. Paul, F., South, P. F., Cavanagh, A. P., et al., Synthetic glycolate metabolism pathways stimulate crop growth and productivity in the field. *Science*, 2019, Vol. 363.

86. Tennyson, A., In Memoriam A. H. H., Edward Moxon, 1850.

87. World Resources Institute, How to sustainably feed 10 billion people by 2050 in 21 charts, 2018. https://www .wri.org/blog/2018/12/how-sustainably-feed-10-billion-people-2050-21-charts.

88. Sullivan M. J. P., Lewis S. L., Affum-Baffoe K., et al., Long-term thermal sensitivity of Earth's tropical forests. *Science*, 2020, Vol. 368, pp. 869–74.

89. Malthus, T. R., *An Essay in the Principle of Population or, A view of its past and present effects or Human Happiness*, John Murray, 1798.

90. Ehrlich, P. R., *The Population Bomb*, Sierra Club/ Ballantine Books, 1968.

91. The Best Stats You've Ever Seen, Hans Rosling, 2007. https://www.youtube.com/watch?v=hVimVzgtD6w.

92. World Values Survey Association. http://www .worldvaluessurvey.org/wvs.jsp.

93. Revision of World Population Prospects. UN DESA Population Division, 2019. https://population.un.org/ wpp.

94. Vollset, S. E., Goren E., Yuan C-W., et al., Fertility, mortality, migration, and population scenarios for 195 countries and territories from 2017 to 2100: a forecasting analysis for the Global Burden of Disease Study. *Lancet*, 2020, Vol. 396, pp. 1285–1306.

ACKNOWLEDGMENTS

For my superlative wife. This book and my years in the desert would have never been possible without you. And thank you to my children Olivia, Hanna, Hans Silas, Adam, and Nora. The future belongs to you, and it should.

My family is indulgent, patient, and endlessly supportive. All of you are relentlessly curious about how to make a much kinder and gentler world for all people and all creatures. All illustrations are by precocious and modern Nora Jovine.

Thank you to my aunt, several fathers, mother, brother, in-laws, and the hordes of amazing Londoners that are the distillate of the best the world has to offer. I have been blessed to know the loveliest of people who are fundamentally optimistic. We live near Borough Market, and if you ever want to taste, smell, and see the vibrant colors produced by photosynthesis, just come to this thousand-year-running vegetable market in South London. It makes your heart sing.

I am indebted to my brilliant and equally patient mentors, professors, and guides at Yale, MIT, UCSB, and WHOI. The real miracle is that they turned this recalcitrant, mountain-boy truant, who hid from his teachers in the woods, into a marine biologist. I am equally grateful to all my professional colleagues, generous employers, and long-suffering bosses. They have taught this dreamer to dream even bigger.

A special thanks to the efficient SBB at the Zürich airport station. I lost the only copy of this manuscript on a Swiss train with my passports and tickets. Yet, in seventy-five minutes, all was returned in perfect condition.

INDEX

ABOUT THE AUTHOR

RAFFAEL JOVINE trained in molecular biophysics and biochemistry at Yale, did his PhD in marine sciences at UC Santa Barbara, and completed research at MIT. In 2013, he founded and is now chief scientist for a company that uses seawater, sunlight, and wind to grow food in coastal deserts, replicating algal blooms. He is married with five children and lives in London.